MOBILE AND WIRELESS COMMUNICATION

AUTHOR (S)

Dr. Maulin Joshi
Professor, Sarvajanik University, Surat

Dr. Urvashi P. Shukla
Assistant Professor, Banasthali Vidyapith, Newai

REVIEWER

Dr. Sandeep Kumar Amera
Assistant Professor, National Institute of Technology, Surathkal

KHANNA BOOK PUBLISHING CO. (P) Ltd.
4C/4344, Ansari Road, Darya Ganj, New Delhi, India

All India Council for Technical Education
Nelson Mandela Marg, Vasant Kunj, New Delhi, India

Dear Readers,

To prevent the piracy, this book is secured with HIGH SECURITY HOLOGRAM on the front title cover. In case you don't find the hologram on the front cover title, please write us to at contact@khannabooks.com or whatsapp us at +91-99109 09320 and avail special gift voucher for yourself.

Specimen of Hologram on front Cover title:

Moreover, there is a SPECIAL DISCOUNT COUPON for you with EVERY HOLOGRAM.

How to avail this SPECIAL DISCOUNT:

Step 1: Scratch the hologram

Step 2: Under the scratch area, your "coupon code" is available

Step 3: Logon to www.khannabooks.com

Step 4: Use your "coupon code" in the shopping cart and get your copy at a special discount

Step 5: Enjoy your reading!

ISBN: 978-93-5538-810-0

Book Code: DIP495EN

Mobile and Wireless Communication

by Dr. Maulin Joshi and Dr. Urvashi P. Shukla

First Edition: 2026

Originally Published by:

All India Council for Technical Education, Ministry of Education, GOI

Exclusively Published & Distributed by:

Khanna Book Publishing Co. (P) Ltd.
4C/4344, Ansari Road, Daryaganj,
New Delhi (INDIA)

Visit us at: **www.khannabooks.com**
Write us at: **contact@khannabooks.com**

To view complete list of books, Please scan the QR Code: KPH

Printed in Bharat.

प्रो. टी. जी. सीताराम
अध्यक्ष
Prof. T. G. Sitharam
Chairman

सत्यमेव जयते

आज़ादी का अमृत महोत्सव

अखिल भारतीय तकनीकी शिक्षा परिषद्
(भारत सरकार का एक सांविधिक निकाय)
(शिक्षा मंत्रालय, भारत सरकार)
नेल्सन मंडेला मार्ग, वसंत कुंज, नई दिल्ली–110070
दूरभाष : 011–26131498
ई–मेल : chairman@aicte-india.org

ALL INDIA COUNCIL FOR TECHNICAL EDUCATION
(A STATUTORY BODY OF THE GOVT. OF INDIA)
(Ministry of Education, Govt. of India)
Nelson Mandela Marg, Vasant Kunj, New Delhi-110070
Phone : 011-26131498
E-mail : chairman@aicte-india.org

FOREWORD

Engineers are the backbone of any modern society. They are the ones responsible for the marvels as well as the improved quality of life across the world. Engineers have driven humanity towards greater heights in a more evolved and unprecedented manner.

The All India Council for Technical Education (AICTE), have spared no efforts towards the strengthening of the technical education in the country. AICTE is always committed towards promoting quality Technical Education to make India a modern developed nation emphasizing on the overall welfare of mankind.

An array of initiatives has been taken by AICTE in last decade which have been accelerated now by the National Education Policy (NEP) 2020. The implementation of NEP under the visionary leadership of Hon'ble Prime Minister of India envisages the provision for education in regional languages to all, thereby ensuring that every graduate becomes competent enough and is in a position to contribute towards the national growth and development through innovation & entrepreneurship.

One of the spheres where AICTE had been relentlessly working since past couple of years is providing high quality original technical contents at Under Graduate & Diploma level prepared and translated by eminent educators in various Indian languages to its aspirants. For students pursuing 2nd year of their Engineering education, AICTE has identified 87 books, which shall be translated into 12 Indian languages - Hindi, Tamil, Gujarati, Odia, Bengali, Kannada, Urdu, Punjabi, Telugu, Marathi, Assamese & Malayalam. In addition to the English medium, books in different Indian Languages are going to support the students to understand the concepts in their respective mother tongue.

On behalf of AICTE, I express sincere gratitude to all distinguished authors, reviewers and translators from the renowned institutions of high repute for their admirable contribution in a record span of time.

AICTE is confident that these outcomes based original contents shall help aspirants to master the subject with comprehension and greater ease.

(Prof. T. G. Sitharam)

ACKNOWLEDGEMENT

The authors are grateful to the authorities of AICTE, particularly Prof. (Dr.) T G Sitharam, Chairman; Dr. Abhay Jere, Vice-Chairman, Prof. Rajive Kumar, Member-Secretary, Dr. Sunil Luthra, Director and Reena Sharma, Hindi Officer Training and Learning Bureau for their planning to publish the books on Computer System Organization. We sincerely acknowledge the valuable contributions of the reviewer of the book Dr. Sandeep Kumar Amera, Assistant Professor, National Institute of Technology, Surathkal. We are very much thankful to Dr. Tejal Joshi, Sarvajanik University, Surat for editing help extended for timely completion of book.

This book is an outcome of various suggestions of AICTE members, experts and authors who shared their opinion and thought to further develop engineering education in our country. Acknowledgements are due to the contributors and different workers in this field whose published books, review articles, papers, photographs, footnotes, references and other valuable information enriched us at the time of writing the book.

Dr. Maulin Joshi

Dr. Urvashi Prakash Shukla

PREFACE

Reflecting vast experience of teaching in the classroom, this text book on "Mobile and Wireless Communication" is written with a view to provide a solid foundation of the subject. Using simple and lucid language, this text book aims to take the reader on a journey to the important technical concepts of mobile wireless communication that is reflected in using one of the most powerful devices Mobile phones-being used in daily life and offices. This book is suitable for use as one-semester course material for diploma students of Electronics, Electronics and communication engineering, Computer and IT engineering, Electrical and electronics engineering, Electronics & Instrumentation engineering.

This book is organised in total 4 modules. The outline of the book is as follows:

First unit describes an exploration of the evolution of wireless communication, spanning its historical origins to the latest 5G technology. It spans over mobile radio systems, including mainly cellular mobile systems plus cordless telephony, with real-world examples for practical insight. The unit also focuses on the fundamentals of system design, addressing frequency reuse, channel assignment, handoff types, and managing interference and system capacity. This equips students with essential knowledge for designing robust wireless communication systems.

Second unit provides in-depth knowledge of wireless propagation, covering many areas necessary to understand and enhance wireless communication systems. It also covers topics such as link budget, that accounts for transmitter power, antenna gains, and losses when calculating signal strength. In addition, the unit also examines free-space route loss, which explains why signals in open areas deteriorate over extended distances. It discusses multipath fading, shadowing, fading margin, and shadowing margin of the receiver, providing tips and roadblocks to lessen signal deterioration.

Third unit of this book offers a comprehensive examination of the imperative role diversity plays in modern wireless communication systems. Beginning with an exploration of the compelling need for diversity, readers will gain insight into the fundamental reasons behind its incorporation in communication infrastructure. Delving into the various types of diversity, including spatial, frequency, time, and polarization diversity, readers will develop a nuanced understanding of the multifaceted approaches to enhancing signal reliability and performance. In addition, a significant focus of the unit is directed towards space diversity techniques, encompassing Selection Combining, Maximum Ratio Combining, Equal Gain Combining,

Square Law Combining, and Minimum Input Minimum Output (MIMO) antenna systems. Through detailed explanations and illustrative examples, readers will grasp the operational principles and applications of each technique, thus cultivating proficiency in selecting and implementing appropriate diversity schemes. Moreover, this module addresses the pivotal aspect of antenna systems' capacity within the context of space diversity. By elucidating concepts such as antenna gain, radiation patterns, and spatial multiplexing, students will acquire the foundational knowledge essential for optimizing system capacity and spectral efficiency.

In the last couple of decades with the increased demands of mobile wireless systems, issues of limited spectrum efficiency and multipath propagation have also increased. To tackle such issues, fourth unit starts with spread spectrum techniques, and their characteristics, types including direct sequence spread spectrum, and frequency hopped Spread Spectrum. Intending to increase wireless network capacity and provide a good alternative to existing time and frequency-based multiple access techniques, the unit also builds basic knowledge about CDMA, its types, channels, call processing etc. It covers a discussion about the capacity of CDMA channels and a comparison with GSM. The module also presents the concept of orthogonality and examines the suitability of OFDM *as a* multi-carrier system. At last, this unit discusses the evolution of LTE systems that will be helpful as an essential tool for designing and maintaining current-generation mobile wireless systems.

Dr. Maulin Joshi

Dr. Urvashi Prakash Shukla

OUTCOME BASED EDUCATION

For the implementation of an outcome based education the first requirement is to develop an outcome based curriculum and incorporate an outcome based assessment in the education system. By going through outcome based assessments evaluators will be able to evaluate whether the students have achieved the outlined standard, specific and measurable outcomes. With the proper incorporation of outcome based education there will be a definite commitment to achieve a minimum standard for all learners without giving up at any level. At the end of the programme running with the aid of outcome based education, a student will be able to arrive at the following outcomes:

PO1. **Basic and Discipline specific knowledge:** Apply knowledge of basic mathematics, science and engineering fundamentals and engineering specialization to solve the engineering problems.

PO2. **Problem analysis:** Identify and analyses well-defined engineering problems using codified standard methods.

PO3. **Design/development of solutions:** Design solutions for well-defined technical problems and assist with the design of systems components or processes to meet specified needs.

PO4. **Engineering Tools, Experimentation and Testing:** Apply modern engineering tools and appropriate technique to conduct standard tests and measurements.

PO5. **Engineering practices for society, sustainability and environment:** Apply appropriate technology in context of society, sustainability, environment and ethical practices.

PO6. **Project Management:** Use engineering management principles individually, as a team member or a leader to manage projects and effectively communicate about well-defined engineering activities.

PO7. **Life-long learning:** Ability to analyse individual needs and engage in updating in the context of technological changes.

COURSE OUTCOMES

After completion of the course, the students will be able to:

CO-1: Discuss fundamentals of wireless communication and various generations of mobile communications

CO-2: Design parameters of a basic cellular network and apply them for capacity enhancement

CO-3: Describe wireless propagation, link budget and related evaluation parameters for wireless and mobile communication

CO-4: Describe diversity issues of radio propagation models in mobile communications

CO-5: Describe overview of concepts of CDMA , OFDM and LTE

Mapping of Course Outcomes with Programme Outcomes to be done according to the matrix given below:

Course Outcomes	Expected Mapping with Programme Outcomes *(1- Weak Correlation; 2- Medium correlation; 3- Strong Correlation)*						
	PO-1	**PO-2**	**PO-3**	**PO-4**	**PO-5**	**PO-6**	**PO-7**
CO-1	3	2	2	1	2	1	3
CO-2	3	3	3	3	2	2	3
CO-3	3	3	2	3	2	2	3
CO-4	3	3	2	2	1	2	1
CO-5	3	2	1	2	1	1	2

GUIDELINES FOR TEACHERS

To implement Outcome Based Education (OBE) the knowledge level and skill set of the students should be enhanced. Teachers should take a major responsibility for the proper implementation of OBE. Some of the responsibilities (not limited to) for the teachers in OBE system may be as follows:

- Within reasonable constraints, they should manoeuvre time to the best advantage of all students.
- They should assess the students only upon certain defined criterion without considering any other potential ineligibility to discriminate them.
- They should try to grow the learning abilities of the students to a certain level before they leave the institute.
- They should try to ensure that all the students are equipped with quality knowledge as well as competence after they finish their education.
- They should always encourage the students to develop their ultimate performance capabilities.
- They should facilitate and encourage group work and team work to consolidate newer approaches.
- They should follow Bloom's taxonomy in every part of the assessment.

Bloom's Taxonomy

Level	Teacher should Check	Student should be able to	Possible Mode of Assessment
Create	Students ability to create	Design or Create	Mini project
Evaluate	Students ability to justify	Argue or Defend	Assignment
Analyse	Students ability to distinguish	Differentiate or Distinguish	Project/Lab Methodology
Apply	Students ability to use information	Operate or Demonstrate	Technical Presentation/ Demonstration
Understand	Students ability to explain the ideas	Explain or Classify	Presentation/Seminar
Remember	Students ability to recall (or remember)	Define or Recall	Quiz

GUIDELINES FOR STUDENTS

Students should take equal responsibility for implementing the OBE. Some of the responsibilities (not limited to) for the students in OBE system are as follows:

- Students should be well aware of each UO before the start of a unit in each and every course.
- Students should be well aware of each CO before the start of the course.
- Students should be well aware of each PO before the start of the programme.
- Students should think critically and reasonably with proper reflection and action.
- Learning of the students should be connected and integrated with practical and real life consequences.
- Students should be well aware of their competency at every level of OBE.

LIST OF ABBREVIATIONS

General Terms

Abbreviations	Full form
1G	First Generation
2G	Second Generation
3G	Third Generation
4G	Fourth Generation
5G	Fifth Generation
ACH	Adjacent channel
ACI	Adjacent channel Interference
AMPS	Advanced Mobile Phones Services
AWGN	Additive White Gaussian Noise
BER	Bit Error Rate
CA	Channel Assignment
CAI	Co-Antenna Interference
CBRS	Citizens Broadband Radio Service
CCI	Co-channel Interference
CDMA	Code Division Multiple Access
CIR	Channel Impulse Response
CSI	Channel State Information
CSS	Chirped Spread Spectrum
CTS	Continuous-Time Systems
DS	Direct-Sequence
DS	Direct sequence
DSSS	Direct Sequence Spread Spectrum
DTS	Discrete-Time Systems
EDGE	Enhance Data Rates Global Evolution
EGC	Equal-Gain Combining
EIRP	Effective Isotropic Radiated Power
MIMO	Multiple Input Multiple Output
MLSR	Maximum Length Shift Registers
MRC	Maximal-Ratio Combining

Abbreviations	Full form
ERP	Effective Radiated Power
FAF	Floor Attenuation Factor
FDMA	Frequency Division Multiple Access
FFT	Fourier Transform
FHSS	Frequency Hopping Spread Spectrum
FM	Frequency Modulation
FRR	Frequency Re-use Ratio
FSK	Frequency Shift Keying
GMSK	Gaussian Minimum Shift Keying
GPRS	General Packet Radio Systems
GSA	Global Mobile Suppliers Association
GSM	Global System for Mobile
HF	High-Frequency
HPBW	Half-Power Beam Width
HSDA	High-Speed Downlink Access
HSDPA	High-speed Downlink Packet Access
IMT	International Mobile Telecom.
IoT	Internet of Things
ISI	Inter-Symbol Interference
LF	Low Frequency
LFSR	Linear Feedback Shift Register
LMDS	Local multipoint distribution service
LOS	Line of Sight
LTE	Long Term Evolution
MAHO	Mobile Assisted Handoff
MF	Medium Frequency
MISO	Multiple Input Single Output
SS	Spread Spectrum
STBC	Space-Time Block Coding

General Terms

Abbreviations	Full form
MSE	Mean Square Error
MTMR	Multiple-Tx, Multiple-Rx
NLOS	Non-Line of Sight
OFDM	Orthogonal Freq. Div Multiplexing
PDC	Personal Digital Communications
PN	Pseudo Noise
PPM	Pulse Position Modulation
PSD	Power Spectral Density
QOS	Quality of Service
RMS	Root mean square
SC	Selection Combining
SDMA	Space Division Multiple Access
SIMO	Single Input Multiple Output
SIR	Signal-to-interference ratio
SISO	Single Input Single Output
SLC	Square-Law Combining
SNR	Signal-to-Noise Ratio

Abbreviations	Full form
TCP/IP	Transmission Control Protocol/ Internet Protocol
TDD	Time Division Duplex
TDMA	Time Division Multiple Access
THSS	Time-Hopping Spread Spectrum
TIV	Time-invariant
T-R	Transmitter-Receiver
TV	Time-Varying
UBW	Ultra-Wide band
UMTS	Universal Mobile Telecommunications System
VLF	Very Low Frequency
VoIP	Voice over IP
VoLTE	Voice over LTE (VoLTE)
WDMA	Wavelength. Div. Multiple Access
WiMAX	Worldwide Interoperability for Microwave Access

LIST OF FIGURES

Unit 1: Cellular Systems: Overview and Fundamentals of Design Concepts

Figure 1.1 Conceptual diagram of Control and Traffic Channels between BS and MS 5

Figure 1.2 Concept of a) FDD Channels and b) TDD Channels 5

Figure 1.3 Schematic diagram of a Paging System 7

Figure 1.4 A Cordless Telephone System 8

Figure 1.5 Conceptual Diagram of a Cellular System 9

Figure 1.6 Conceptual diagram of the FDMA scheme 12

Figure 1.7 Conceptual diagram of TDMA scheme 14

Figure 1.8 Conceptual diagram for CDMA 15

Figure 1.9 Summary of three multiple access techniques 16

Figure 1.10 Conceptual diagram of Cellular Frequency Reuse. 24

Figure 1.11 Co-channel (CCH) cell location found in a cellular 24

Figure 1.12 Conceptual diagram showing need of Handoff mechanism 12

Figure 1.13 Conceptual diagram showing Co-channel (Co-H) cells 13

Figure 1.14 cellular size N=7, the first tier of co-channel cells 33

Figure 1.15 Conceptual diagrams to illustrate Near-Far problem 35

Figure 1.16 Cell sectioning of various geometries. 37

Figure 1.17 Illustration of cell-splitting 38

Unit 2: Mobile Wireless Propagation

Figure 2.1 Thévenin model of a resistor. 56

Figure 2.2 White noise spectral model. 58

Figure 2.3 Noise Model of amplifier 59

Figure 2.4 Two-port model for a communications system component. 60

Figure 2.5 Calculation of noise for cascaded system 61

Figure 2.6 Geometry for calculating the reflection coefficients between two dielectrics. 63

Figure 2.7 Two-ray ground reflection model 65

Figure 2.8 Illustration of knife-edge diffraction geometry 67

Figure 2.9 Large and small-scale variations in signal strength over time 73
Figure 2.10 PDF of received signal level following Log-normal distribution 76
Figure 2.11 Street Microcell 78
Figure 2.12 Illustrating plane waves arriving at random angles. 88
Unit 3: Diversity, Capacity and MIMO Systems
Figure 3.1 (a) Types of small-scale Fading (Based on Multi Path delay Spread) and (b) Types of small-scale Fading (Based on Doppler Spread) 101
Figure 3.2 Types of Diversity. 104
Figure 3.3 Block diagram of space diversity 107
Figure 3.4 Block diagram of MRC using multiple receive antennas 108
Figure 3.5 Conceptual diagram of different systems along with their spatial diversity 117
Figure 3.6 SIMO system example 118
Figure 3.7 Beamforming (beam steering) with two transmitting antennas. 120
Figure 3.8 Diversity with two receiving antennas 121
Figure 3.9 Combination of Beamforming (beam steering) and Diversity in smart antenna. 121
Figure 3.10 MIMO using Multi Tx- Multi -Rx to send a multi-signal on the same channel 121
Figure 3.11 A simple MIMO wireless transmission system 123
Figure 3.12 Subtractive addition of multipath signal resulting in direct, delayed (slightly attenuated) and added (heavily attenuated) signal 125
Figure 3.13 Subtractive addition of multipath signal resulting in direct and reflected signals with a large difference 125
Figure 3.14 A design of M-M-branch antenna diversity receivers 127
Figure 3.15 Block diagram of MIMO wireless link. 128
Figure 3.16 Basic channel model 129
Unit 4: Overview of CDMA, OFDM and LTE
Figure 4.1 Conceptual diagram of Frequency Hopping Spread Spectrum (FHSS) 141
Figure 4.2 Conceptual diagram of Direct sequence Spread Spectrum (DSSS). 142
Figure 4.3 Illustration of transmission of '10' code (1 followed by 0) using DSSS 143

Figure 4.4 LFSR using Galois implementation and modulo two sum 148

Figure 4.5 Carrier modulation and spreading of original data to get SSM signal. 151

Figure 4.6 Carrier demodulation and dispreading of SSM signal to get the original data. 151

Figure 4.7 Near Far Problem Scenario 152

Figure 4.8 Illustration of Closed and Open Loop Power Control System in Mobile 153

Figure 4.9 Single Carrier Allocation of Frequency to the Users in a network. 155

Figure 4.10 Application of Windowing Function to the OFDM baseband Signal. 159

LIST OF TABLES

Table 1.1 Comparison of 2G, 3G, 4G and 5G 20

Table 1.2 Co-channel Reuse Ratio for Some Values of N 30

Table 2.1 Comparison of Macro Cell vs. Microcell 79

Table 3.1 Countermeasures for Small-scale fading 102

Table 3.2 Types of Diversity System 117

Table 3.3 Comparison between the Smart Antenna and MIMO. 122

CONTENTS

Foreword iii
Acknowledgement iv
Preface v
Outcome Based Education vii
Course Outcomes x
Guidelines for Teachers ix
Guidelines for Students x
Abbreviations and Symbols xi
List of Figures xiii
List of Tables xv

Unit 1: Cellular Systems: Overview and Fundamentals of Design Concepts ***1-44***
Unit specifics 1
Rationale 1
Pre-requisites 2
Unit outcomes 2
1.1 History of Wireless Communication 2
1.1.1 Basic Terminologies for Wireless Communications System 3
1.2. Examples of Mobile Radio Systems 5
1.2.1. Paging System 6
1.2.2. Cordless Systems 7
1.2.3. Cellular Mobile Systems 8
1.2.3.1. How a cellular call is made 9
1.3. Evolution of Cellular Wireless Systems 11
1.3.1. Basics of Multiple Access Schemes 11
1.3.2. Evolution (or Transformation) of Mobile Cellular Networks 16
1.4. Cellular concepts - fundamentals of system design 21
1.4.1. Introduction 21
1.4.2. Frequency reuse 22
1.4.3. Channel Assignment (CA) Types 25
1.4.4. Types of Handoff 26
1.4.5. Interference and System Capacity 28
1.4.5.1. Co-channel interference (CCI) 28

1.4.5.2. Adjacent channel interference (ACI) 35
1.4.5.3. Cell sectorization 36
1.4.5.4. Cell splitting 38
Unit Summary 39

Unit 2: Mobile Wireless Propagation **45-96**
Unit specifics 45
Rationale 45
Pre-requisites 46
Unit outcomes 46
2.1. Highlights of Mobile Wireless Propagations 47
2.2. Noise and Interference 55
2.2.1. Thermal Noise 55
2.2.2. Equivalent Noise Temperature and Noise Figure 58
2.2.3. Noise in Cascaded Systems 59
2.2.4. Man-Made Noise 61
2.3. Three Basic Propagation Mechanisms 62
2.3.1. Reflection 62
2.3.1.1. Reflection from dielectric 63
2.3.1.2. Ground reflection (2-ray) model 64
2.3.2. Diffraction 65
2.3.3. Scattering 68
2.3.4. Some Scenarios of Radio Mechanisms 69
2.4. Radio Propagation Model 70
2.4.1. Introduction 70
2.4.2. Free Space Propagation Model 70
2.4.3. Introduction to Radio Propagation Models 71
2.4.4. Small-Scale Propagation Model 72
2.4.5. Large-Scale Propagation Model 73
2.4.6. Log-Distance Path Loss Model 74
2.4.7. Log Normal Shadowing Model 74
2.4.8. Log Normal Distribution Model 76
2.4.9. Percentage Of Coverage Area 77
2.5. Outdoor Propagation Models 77
2.5.1. Longley Rice Model 79

2.5.2. Okumura Model 80
2.5.3. Hata Model 81
2.6. Indoor propagation models 82
2.6.1. Noise Figure for Link Budget 84
2.7. Multi Path Fading Shadowing, Fading Margin, Shadowing Margin 86
2.8. Statistical Models for Multipath Fading Channels 88
Unit Summary 89

Unit 3: Diversity, Capacity and MIMO Systems ***97-136***
Unit specifics 97
Rationale 97
Pre-requisites 98
Unit outcomes 98
3.1 Introduction to Equalization and Diversity 99
3.1.1 RMS Delay Spread 99
3.1.2 Coherence Bandwidth (BC) and Coherence Time (TC) 99
3.1.3 Equalization and Adaptive Equalization 103
3.2 Types of Diversity 104
3.3 Space diversity techniques 106
3.3.1 Selection Combining (SC) 107
3.3.2 Maximal-Ratio Combining (MRC) 108
3.3.3 Equal-Gain Combining (EGC) 109
3.3.4 Square-Law Combining (SLC) 110
3.4 Wireless Channel Capacity 110
3.4.1 Capacity Measurements In AWGN Channel 111
3.4.2 Capacity of Flat-Fading Channels 113
3.4.3 Capacity With Receiver Diversity 113
3.4.4 Capacity of Frequency-Selective Fading Channels 114
3.5 MIMO 115
3.5.1 Introduction 115
3.5.2 Space diversity: concept and system 116
3.5.3 Smart Antenna and MIMO 119
3.5.4 MIMO-Based Architecture 123
3.5.5 MIMO- a Solution for Multipath 124
3.5.6 Multipath Diversity Reception in Multi-Antenna Receiver 126
3.5.7 MIMO Channel Capacity 126

3.5.8 Basic Channel Model 128
3.5.9 Outage Capacity 129
3.5.10 Advantages of MIMO 130
3.5.11 MIMO Applications 130
Unit Summary 131

Unit 4: Overview of CDMA, OFDM and LTE ***137-169***
Unit specifics 137
Rationale 137
Pre-requisites 138
Unit outcomes 138
4.1 Introduction to CDMA signal generation 138
4.1.1. Spread Spectrum Modulation 139
4.1.2. Types of Spread Spectrum Technique 140
4.1.2.1.Frequency hopping spread spectrum (FHSS) 140
4.1.2.2.Advantages and disadvantages of FHSS 141
4.1.2.3.Direct sequence spread spectrum (DSSS) 142
4.1.2.4.Advantages and disadvantages of DSSS 144
4.1.2.5.Relation between spread spectrum bandwidth and signal to noise ratio 144
4.1.3. Types Of Codes Used In CDMA Signal Generation 145
4.1.3.1.Walsh code 145
4.1.3.2.Pseudo noise (PN) codes or direct sequence (DS) fundamentals 146
4.1.3.3.Properties of PN codes 146
4.1.3.4.Maximum length (ML) sequences 147
4.1.4. Capacity of CDMA System 148
4.1.5. Comparison of CDMA and GSM Capacity Trade-offs 150
4.1.6. General Block Diagram of DS Transmitter and Receiver 150
4.1.7. Near-Far Problem and Power Control 152
4.2 Overview of OFDM 154
4.2.1. Basic Principle Of Orthogonality 154
4.2.2. Single Vs. Multi-Carrier Systems 154
4.2.2.1.Single-carrier systems 155
4.2.2.2.Multicarrier systems 156
4.2.3. OFDM Signal Mathematical Representation 156

4.2.4. Selection Parameters for Modulation 157
4.2.5. Windowing in OFDM and Spectral Efficiency 159
4.3 Overview of Long-Term Evolution (LTE) 160
4.3.1. Introduction 160
4.3.2. Working of LTE 160
4.3.3. Benefits of LTE 163
4.3.4. Features of 4G LTE 163
4.3.5. Major Milestones In LTE's Development 165
Unit Summary 166
CO-PO Attainable Table 170
Index 171

1 Cellular Systems: Overview and Fundamentals of Design Concepts

UNIT SPECIFICS

Through this unit, the following aspects are discussed:

- *History of wireless communication*
- *Mobile Radio Systems*
- *Evolution of cellular wireless systems*
- *Fundamentals of system design*

RATIONALE

This module provides a concise exploration of the evolution of wireless communication, spanning its historical origins to the latest 5G technology. It covers mobile radio systems, including paging, cordless telephony, and cellular mobile systems, with real-world examples for practical insight. The module then focuses on the fundamentals of system design, addressing frequency reuse, channel assignment, handoff types, and managing interference and system capacity. These equip students with essential knowledge for designing robust wireless communication systems. By the end, students will have a comprehensive understanding of the historical context, technological advancements, and fundamental principles necessary for navigating the dynamic field of wireless communication.

PRE-REQUISITES

Basics of Signal and Wave

Basics of Communication systems

UNIT OUTCOMES

Following outcomes are expected after going through this unit:

U1-O1: Describe the basic components of a Mobile Wireless Communications System

U1-O2: Explain basic multiple access techniques used in mobile Wireless Communications System

U1-O3: Describe various generation Evolution of mobile cellular networks

U1-O4: Improve basic design of the mobile Wireless Communications System in terms of channel capacity using fundamental components

	MAPPING of UNIT OUTCOMES WITH COURSE OUTCOMES ***(1- Weak Correlation; 2- Medium correlation; 3- Strong Correlation)***				
	CO-1	***CO-2***	***CO-3***	***CO-4***	***CO-5***
U1-O1	*3*	*2*	*2*	*1*	*1*
U1-O2	*3*	*2*	*--*	*2*	*2*
U1-O3	*3*	*2*	*1*	*1*	*3*
U1-O4	*2*	*3*	*3*	*3*	*2*

1.1. History of Wireless Communication

Wireless communication has come a long way since its inception in the mid-19th century. In 1865, Maxwell presented a theory of the existence of electromagnetic waves, which Hertz later proved in 1887 through his experimental transmission and reception of these waves. His findings opened up practical applications for this technology.

Guglielmo Marconi was a key figure in the early 20th century, building upon Hertz's work to transmit transatlantic radio signals in 1901. This led to rapid advancements and the invention of the triode amplifier by Lee De Forest, which enabled voice transmission and the era of radio broadcasting. Wireless communication continued to flourish throughout the 20th century with the development of radar, mobile phones (starting with bulky car phones in the 1940s), and advancements in radio technologies like FM and AM. The invention of the transistor miniaturized electronics, making mobile phones smaller and more portable.

The late 20th and early 21st centuries saw a significant rise in wireless communication with the arrival of cellular networks. The First Generation (1G) provided basic voice calls. This was followed by 2G, which introduced SMS and data capabilities. 3G brought mobile internet, and 4G transformed speeds, paving the way for the smartphone revolution. Today, 5G is on the horizon, promising even faster speeds and lower latency, transforming everything from mobile gaming to autonomous vehicles.

1.1.1. Basic Terminologies for Wireless Communications System

- **Base Station (BS):** In general, a fixed location is used to communicate with mobile stations. BS can be located at the centre of a cell or on the edge of a cell and it is equipped with various channels, transmitter, and receiver equipment.
- **Cell:** The division of the area of coverage is usually called a cell. Each cell has its base station usually located at its centre or sometimes at the edge.
- **Control Channel (CCH):** A radio channel dedicated to transmitting call setup, call request or call initiation. Figure 1.1: shows the concept of Control Channels existing between BS and Mobile station (MS).
- **Forward Channel (downlink):** A radio channel that transmits information from the BS to the MS.

- **Frequency Division Duplex (FDD):** both the BS and MS units transmit and receive signals simultaneously. At the BS, there can be two separate transmit and receive antennas. At the MS, however, only a single-keeping antenna saves space and cost. Additionally, a device called a duplexer facilitates common antenna usage for transmission as well as reception of signals.
- **Full Duplex Systems:** These systems support bidirectional communication having independent channels for transmission and reception.
- **Half Duplex Systems:** These types of systems also support bidirectional communication on a given channel but only either send or receive at a time.
- **Handoff:** Method of transferring controls of given MS from one BS to another BS.
- **Mobile Station (MS):** A mobile equipment designed for use while moving.
- **Mobile Switching Center (MSC):** An important unit that handles call routing in its area, connecting cellular BS and mobiles and also to the Public Switched Telephone Network (PSTN).
- **Page:** A message broadcasted over the complete coverage area by multiple BSs at once. Primarily this is used to locate desired MS in a geographic area.
- **Reverse Channel (uplink):** A radio channel dedicated to transmitting messages from the MS to BS.
- **Simplex Systems:** These systems support only unidirectional communication.
- **Subscriber:** A user who subscribes to services by paying required charges for using a mobile communications system.
- **Time Division Duplex (TDD)**: using TDD, on a single channel, at a time instant time (say $t1$) transmission from the BS to MS and some other time (say $t2$) transmission from the MS to BS can be done.
- **Traffic Channel:** channel responsible for carrying (transfer) of data between BS and MS.
- **Transceiver:** A radio signal transmitter and receiver that operates simultaneously.

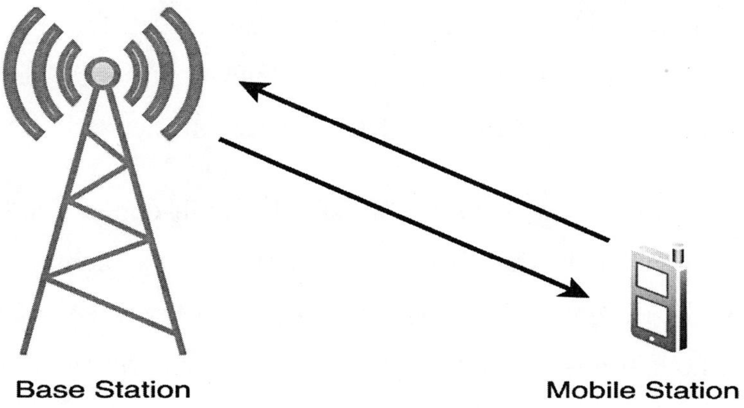

Figure 1.1 Conceptual diagram of Control and Traffic Channels between BS and MS

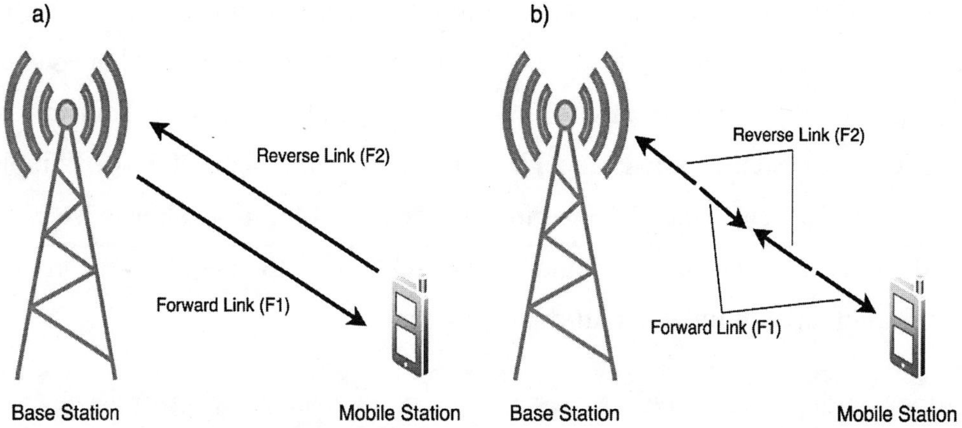

Figure 1.2 Concept of a) FDD Channels and b) TDD Channels

1.2. Examples of Mobile Radio Systems

Mobile radio communication systems are all around us, silently orchestrating many of our daily interactions. From staying connected on smartphones to navigating with GPS, our lives are woven with mobile radio communication systems.

The word "mobile" has traditionally been applied to a radio device that can be moved during operation, whereas "portable" refers to a hand-held radio terminal. The term "subscriber" refers to a user of mobile or portable devices who pays a subscription fee to utilize the system. In a wireless system, the collective group of users is known as "users" or "mobiles," and they communicate with fixed BS that is connected to a backbone network.

Mobile radio transmission systems can be simplex, half-duplex, or full-duplex. Simplex systems are one-directional, while half-duplex allow two-way communication on the same radio channel, with users only able to transmit or receive at any given time. Full-duplex enables simultaneous radio transmission and reception by providing separate channels or appropriate time slots on a single channel.

Frequency Division Duplexing (FDD) enables simultaneous reception and transmission between the subscriber and base station using separate transmit and receive channels. The subscriber unit uses a single antenna and a duplexer to allow for simultaneous transmission and reception. FDD requires a 5% frequency separation between transmit and receive frequencies. The system's radio channel is defined by a combined 'set of two' simplex channels with specified frequency spacing. The forward channel transports signals from the BS to the MS, and the reverse channel transports traffic back to the BS. Transceivers are used for radio communication in both full and half-duplex systems.

Time-division duplexing (TDD) shares a single radio channel by transmitting from the BS to the MS during a portion of time and from the MS to BS during the remaining time. The performance of TDD requires good transmission modulation and timing synchronization. TDD powers Wi-Fi networks, some 4G/5G LTE deployments, and even satellite communication systems. Its dynamic nature and spectral efficiency make it a valuable tool for optimizing communication in an increasingly data-driven world.

1.2.1. Paging Systems

Paging systems, as shown in Figure 1.3, provide brief messages to subscribers in numeric, alphanumeric, or voice format to provide notifications or instructions. Subscribers receive messages, which include news, stock quotations, and faxes, using a toll-free number and a modem or keypad. The system sends out messages via base stations on a radio carrier.

They range in complexity and scope, with simple systems spanning 2-5 km and wide-area systems providing global coverage. Wide area networks simulcast using a network of telephone connections, radio towers and base station transmitters. They provide dependable connections for users in a variety of places by utilizing high transmitter power and modest data rates.

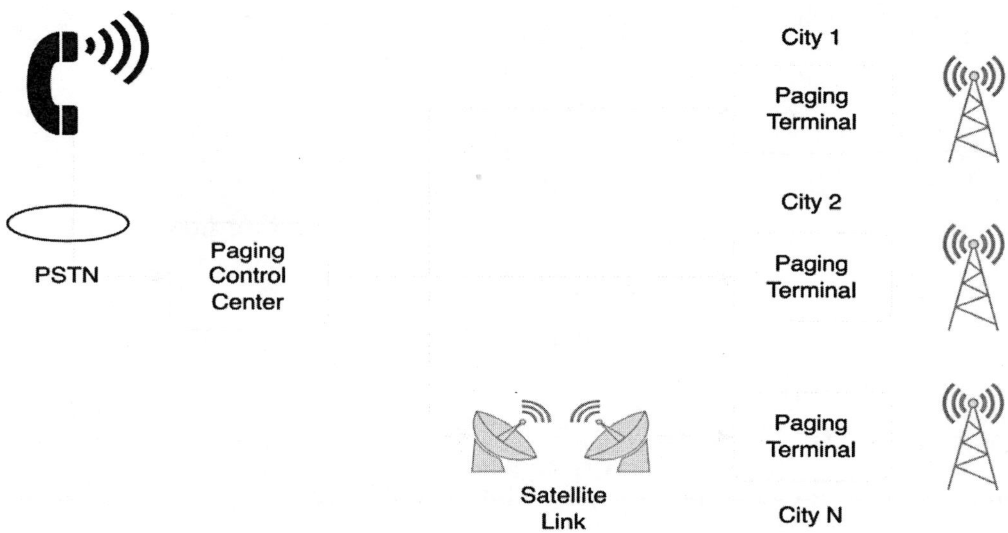

Figure 1.3 Schematic diagram of a Paging System

1.2.2. Cordless Systems

Cordless systems provide FDD communication via radio links between a handset that can move in a limited area and its dedicated BS. Through the PSTN, this base station is connected to a particular phone line that has its number. When cordless phone systems were first introduced in the 1980s, the mobile phone only connected with a specific base unit, usually over short distances. These early cordless phones are mostly meant for home usage, and they operate as extensions of a transceiver that is connected to a PSTN subscriber line.

At places, second-generation cordless phones enabled users to use their devices at various outdoor spots. Paging receivers and modern cordless phones could occasionally be paired so that a subscriber could get page notifications first and reply to them using the cordless phone.

Because a call could not usually be maintained when a user walks outside of the base station's service area, cordless telephone systems had a limited range and range of motion. Typical 2G BSs gave coverage distances of up to a few hundred meters. A cordless telephone system is illustrated in Figure 1.4.

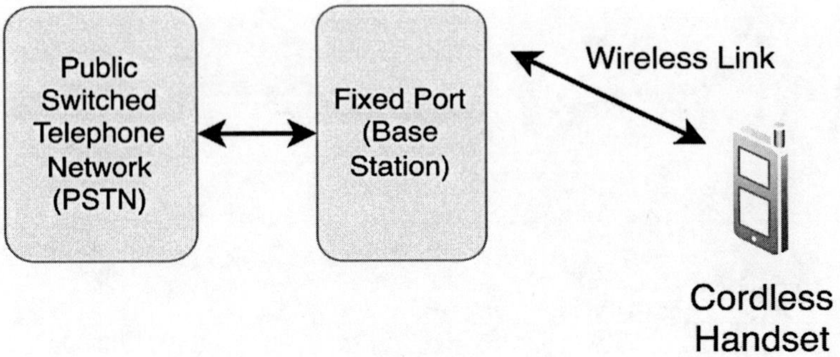

Figure 1.4 A Cordless Telephone System

1.2.3. Cellular Mobile Systems

A cellular telephone system allows users within its radio range to connect wirelessly to the PSTN. These systems use limited frequency resources to accommodate many customers across a large area. Cellular radio systems have emerged as a good alternative to earlier landline telephone networks. Each BS can navigate in a designated geographic area called a cell to maximize capacity, allowing distant base stations to reuse radio channels. Handoff, a complex switching procedure, allows for smooth call transitions across cells. Figure 1.5 depicts a conceptual diagram of the cellular system, which includes MSs, BSs, and a Mobile Switching Centre (MSC). The MSC connects mobile phones to the PSTN. Mobiles communicate with base stations via radio signals, which can be handed off to other base stations during calls. Base stations have transmitters and receivers supporting full duplex communications and connect mobile users in the cell to the MSC. In the cellular system, MSC coordinates base station activities, handles system maintenance, and billing and connects the cellular network to the PSTN. For communication four channels: forward voice channels, reverse voice channels, forward CCHs, and reverse CCHs. CCHs facilitate various control actions including call initiation and channel changes for mobiles.

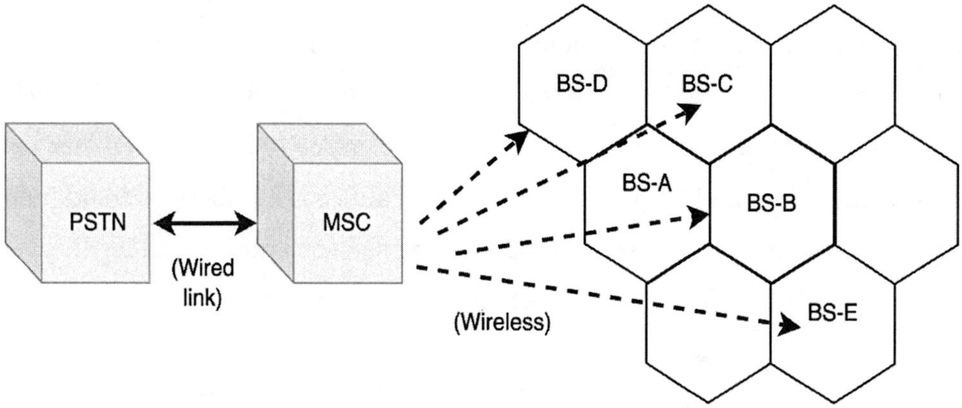

Figure 1.5 Conceptual Diagram of a Cellular System

1.2.3.1. How a cellular call is made

A cell phone uses battery power to scan forward CCHs for the strongest signal. It constantly analyzes signal strength and recognizes at least one or more base stations with which it can communicate. The display may change depending on which base station it is connected to. CCHs are specified and standardized across the service area, accounting for up to 5% of the total number of channels. This means that 5% of your resources go into controlling and setting up your calls, even when you're not making a call.

When you call a mobile user, the mobile switching centre, which is connected to the PSTN (Public Switched Telephone Network), sends a request to all base stations. This can be accomplished using fibre or point-to-point microwave connections. Once the request is received, all base stations receive a paging message including a unique mobile identification number. This number may not be the same as the phone number but is a different code.

The mobile station receives this message because it is already in contact with one base station. When the base station broadcasts this number, the mobile station must respond by identifying itself using the reverse control channel. The BS then sends a handshake to the mobile switching centre, which directs it to switch to an unused voice channel. It's possible that the base station has run out of available voice channels and won't initiate a call. The mobile switching centre employs a lookup table to hold phone numbers and identity numbers, which are needed to ensure legal calls. For example, a phone marked as stolen can be blacklisted, prohibiting calls from being sent. The mobile switching centre normally blacklists the Mobile Identification Number (MIN) number, which remains with the centre. A location register can also store information regarding mobile phones, such as prepaid card numbers, prepaid money, and call charges. Before starting a call, several parameters, such as validity and legality, are evaluated to ensure that the mobile switching centre functions properly.

The GSM concept entails determining whether a response comes from a mobile phone and whether the phone is in good working order with a valid SIM card. This process requires various numbers, including the Subscriber Identity Module (SIM) card, hardware, and logical phone number. The MIN is the hardware number. During a call with a mobile user, the base station instructs the mobile station to switch between unused forward and reverse voice channels. The call cannot be launched until both channels are available. A data message named alert is sent via the forward voice channel to direct the phone to ring. These sequences occur swiftly and are often overlooked by the user.

The mobile switching centre adjusts the transmitted power to maintain call quality. However, modern base stations have more activity taking place at the base station, which cannot handle much traffic and power control issues. As handsets become more microprocessor-friendly, power calculations are also done by the handset itself. The idea is to distribute computing so that the mobile switching centre is not overloaded. In another scenario, a call from a mobile user is made. A call initiation request is sent to the reverse CCH, which transmits the MIN and phone number of the receiver party. These numbers have different utilities in terms of verification at different levels. The mobile switching centre validates the data and makes the connection to the called party through the public switched telephone network, whether it is to a landline number or another mobile.

1.3. Evolution of Cellular Wireless Systems

This section addresses the evolution of current cellular wireless communication systems, as well as why they work, the driving forces behind their operation, and the technological advancements that have been received in these systems.

Since its introduction, there has been a significant increase in the number of cellular subscribers worldwide [1]. It appears that in many developed or undergoing development countries, now older-generation mobile networks are being replaced by new-generation networks. The adoption of 5G is capable of bringing out digital ecosystem development. These include connecting various machines, devices and systems with very low latency and also improving energy efficiency. Improvement in the number of mobile users that started in 1992 with about 23 million cellular subscribers grew steadily and touched the 1 billion mark in 2002. Statistical data reveals that about 40 % of the world's population will now covered by 5G by 2023 and today, access to a mobile-broadband network is available to 95 per cent of the world population. In many countries, older-generation mobile networks are being switched off in favor of new-generation networks. 5G enables the development of a digital ecosystem by connecting machines, objects, and devices with ultra-low latency and the potential to improve energy efficiency. This is the case for most European operators that plan to switch off 3G networks by December 2025 and for operators in the Asia-Pacific region. However, in some countries, the path is less clear, mainly because 2G and 3G networks retain a significant presence. This is the case notably in lower-income countries, where both technologies remain an important means of communication. In these countries, the main obstacles to 5G deployment and adoption include high infrastructure costs, device affordability, and regulatory barriers.

1.3.1. Basics of Multiple Access Schemes

To comprehend this expansion, one must first understand the fundamental concepts behind these cellular communication systems. The multiple-access approach is one of the key contributors to this extraordinary expansion. For technology to serve a high number of users, it must ensure that a sufficient amount of bandwidth is available for voice, text, data, and multimedia communications. Multiple access techniques are typically used to share the limited radio frequency among numerous mobile users. It is vital to keep in mind that radio spectrum is not free and that paying for radio spectrum licensing is one of the biggest expenses. Even with financing available, careful planning and management are necessary to obtain a significant amount of bandwidth. High capacity can only be attained by concurrently assigning the bandwidth, which can be done in several methods including spectrum sharing.

It has to be noted that there are numerous multiple access schemes. Bandwidth as shown earlier, is one of the sources and all the resources are shared in different manners. While sharing resources in mobile communication one constraint has to be kept in mind that there should not be any severe performance degradation in terms of Quality of Service (QoS). QoS varies for different applications. For instance, the maximum delay permitted, the packet loss rate, or the data call drop rate are examples of QoS parameters in voice communication. However, the bit error rate is also a crucial component. We need to make sure there isn't a significant decline in performance while these limitations are in place, even with an increase in the number of users. Frequency Division Multiple Access (FDMA) is one of the most commonly used MA techniques. Figure 1.6 shows the FDMA scheme.

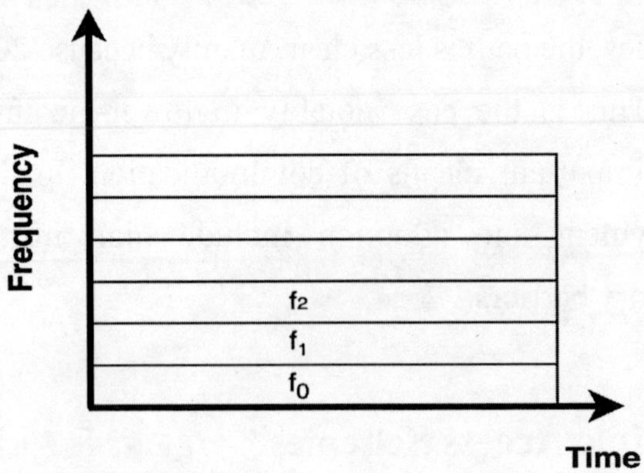

Figure 1.6 Conceptual diagram of the FDMA scheme

Figure 1.6 shows a conceptual diagram of the FDMA scheme where the total bandwidth to be allocated is shown on the y-axis. One should note that available bandwidth is finite. On the x-axis, there is a time. There are three users- user A, user B and user C and three sub-bands namely for f_1 and f_2 are available. These subbands are fixed in terms of bandwidth. When user A decides to utilize services and switches on the phone, the system has to allocate one of the subbands, e.g. subband number f_o is allocated to user A. Suppose, user A is already utilizing services and user 2 switches on the phone and has to be allocated a frequency for talking, we can give them another band (f_1 or f_2) but not necessarily the adjacent band (f_1). Although the subbands have already been determined, the allocation must be done dynamically based on the number of users who plan to communicate after turning on their phones. If a system can create space among different users along the y-axis, it means they have less adjacent channel interference. It should be noted that we need to have ideal filters if closely spaced frequency subbands have to be used in practice.

Now, let us look at an example of how FDMA is used. All three users (A, B and C) are now free to use their frequency bands (f_o, f_1 and f_2) throughout the time domain. All across the time duration, they can use their own allocated entire frequency band without worrying. In a day they can talk, keep quiet, have long periods of silence, etc. They pay for the allocated frequency sub-band. It should be noted that such closely spaced frequency bands have the problem of energy leakage from one band to the other band because of non-ideal filters. These advocates have some kind of a guard band provision. Now the question arises: who decides the width of the guard band? It depends on how sharp the filter cut-offs are. More expensive filters with sharp cut-offs will allow designers to keep narrower guard bands. It is to be noted that guard bands are not used to communicate data.

Another popular multiple access scheme is a Time Division Multiple Access (TDMA) as shown in Figure 1.7. To show a comparison with FDMA, here also frequency (or bandwidth) is shown on the y-axis and time on the x-axis. In this case, in contrast to FDMA (subdivision on frequency axis), now subdivisions are made on the time axis amongst n users (A, B, C and so on). As shown, we have three times slots where user A uses slot 1, user B slot 2 etc. Each user is allowed to utilize the system by speaking or communicating in its time slot. After all time slots are over, time slot 1 will repeat followed by 2, 3 and so and so forth. It should be noted this does not imply that communication will appear broken at the receiver end. This is due to whenever signals are communicated, they are sampled - first at slightly greater than the Nyquist sampling frequency.

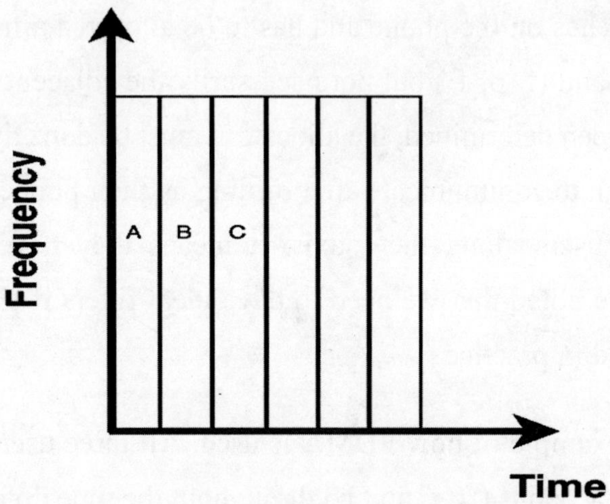

Figure 1.7 Conceptual diagram of TDMA scheme

At the receiver end, reception of a signal at proper sample points is arranged and hence signal reconstruction can be managed. Due to this synchronization mechanism, the signal (e.g. received voice) will not appear broken at the receiver end. Let us try to visualize how it happens in a real-life situation. Consider n users (A, B, C and so on) who are named as mobile phone users and they need to communicate with the base station. So, on the x-axis is the time. User A sends a small packet of information in time slot 1. Similarly, user A sends in time slot 2 and so and so forth. All users will utilize entire time slots 1, 2, 3 ....n in the case of Time Division Multiple Access (TDMA). An important issue to be addressed in TDMA is time synchronization. If time is not synchronized well with the transmitter, one user may likely listen to somebody else's time slot. Hence, the issue of time synchronization is of importance in TDMA. Figure 1.8 shows a conceptual diagram for the third popular multiple access scheme known as Code Division Multiple Access (CDMA).

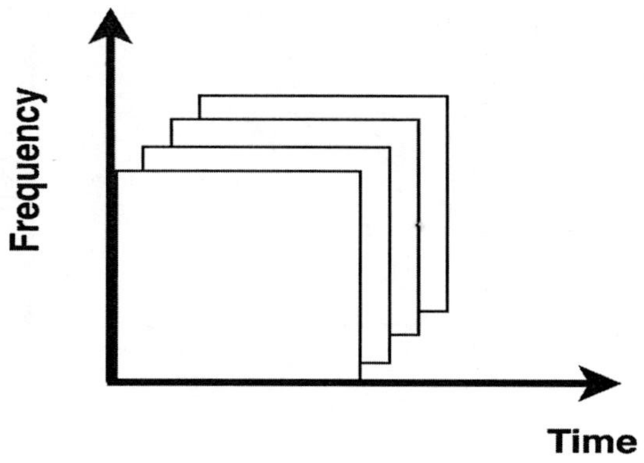

Figure 1. 8 Conceptual diagram for CDMA

In CDMA, one can utilize two popular axes; the frequency on the y-axis and time on the x-axis at the same time i.e. combination of TDMA and FDMA. In this case, every channel consists of a particular time slot and a particular frequency subband. The CDMA scheme adds one more dimension named the code dimension. Due to this, every user is allowed all the time and all the frequency bands for communication but with an assigned particular code. This means that user A will be assigned code 1. User B will have its spreading code and so forth. User n will also have its spreading code. Later it will be shown the reason for calling it a spreading code. At this moment, it is enough to understand that each user can decode the signal using its code. For example, suppose in an international meeting there are ambassadors from n different nations and each ambassador uses a different (own national) language. So one person speaks in English (Code A), one person speaks in Russian (Code B), and one person speaks in Hindi (Code C). There exists a master control called a base station that can understand all these languages. In a given case, if all of the users keep talking in their languages (own codes in CDMA) and the base station who knows all the languages can understand what each speaker is saying. All the users are utilizing entire spectra at the same time. However, as more and more users start speaking, there will be more and more interference. This fact we shall discuss later. The overall conceptual summary of all three multiple-access techniques is shown in Figure 1.9.

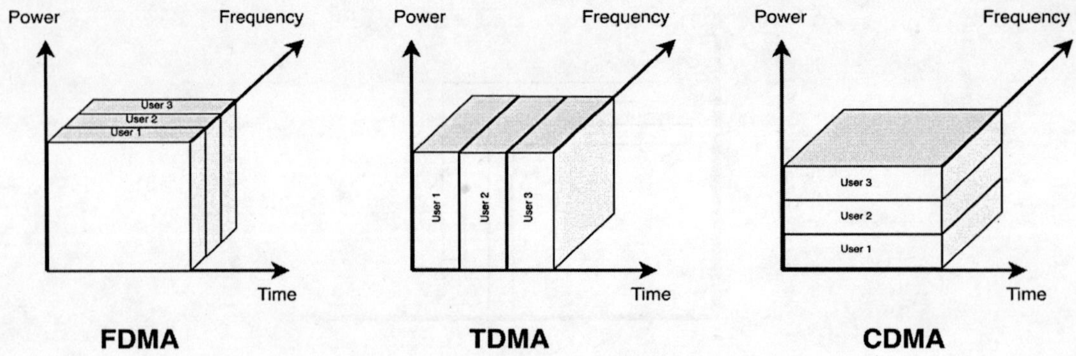

Figure 1. 9 Summary of three multiple access techniques

1.3.2. Evolution (or Transformation) of Mobile Cellular Networks

Let us take an overview of the evolution of cellular networks from the perspective of the multiple access schemes. This is because over the years the system has to cater to the challenge of increasing the number of users and that too with cost-effective mechanisms to generate revenues and propel growth. The journey started with the first generation was launched in the mid-1980s. First-generation mobile cellular networks were purely analogue. They used analogue modulation schemes and mostly Frequency Modulation (FM). Mobile cellular networks were in the primary phase and systems intended primarily for voice traffic because in that era phones meant voice. The system used FDM schemes just to chop up the entire frequency band into subbands and used analogue transmission in each subband. It should be noted that due to a lack of global standardization, first-generation schemes were confined to national boundaries only. An example of such a scheme is the Advanced Mobile Phones Services (AMPS) that was popular in the US. As systems were confined to a local territory services, had to be limited in the number of users. As cellular networks evolved, the second generation also came. This was also developed for voice communication again as it was not thought of using phones for data communication. It was a time when mostly no one believed that it was possible to access the internet through the phone. The Internet itself was in its earlier stages.

I. Second Generation (2G)

It should be noted that, unlike the first-generation, second-generation systems used digital modulations. Global System for Mobile (GSM) started the 2G phone system by using GMSK- the Gaussian Minimum Shift Keying (GMSK) as a digital modulation technique. Second-generation systems used TDMA, FDMA and CDMA as multiple access schemes. Such systems could provide low data rates i.e. order of about 10 kilobits per second. It should be noted that even though they were designed for voice, there was an inherent provision to handle data simply because they were digital schemes. Second-generation GSM used a combination of TDMA and FDMA. 2G networks initially started with 900 MHz and later adopted 1800 MHz as a variant. Another popular 2G system was Personal Digital Communications (PDC) in Japan. IS -95 was different. It was CDMA based on the code division multiple access and Qualcomm the company which essentially proposed CDMA was fueling this growth. It was popular in the US and South Korea. However, 2G had some impediments. It was not suitable for internet services as the original GSM phone was built for a circuit-switched system and could not have a packet-switched service. However, internet services require a packet-switched network. Another general problem with all second-generation systems was they had too many standards. Although GSM took the major chunk of the pie, there were too many standards, one in the US, European standard, Japanese standard etc. So, it was a mess. All these motivated toward the 2.5 G. It resulted in to removal of 2G systems. Some key limitations of 2G can be summarized as below:

- Developed for voice communication (not suitable for data traffic)
- Provided better quality voice calls (compared to 1G) but data speeds of up to order of 10s of kbps
- Not suitable for internet
- Developed with local standards – lacking Global Acceptance

II. Generation (2.5g)

After the successful implementation phase for 2G systems, around 2002 the research community realized that data traffic also should gain momentum and hence technological improvements should be addressed to increase data carrying capability. However, at that time around two-thirds of the mobile phones in the world were GSM. System operators decided to have an add-on called 2.5 G. Some key features of 2.5 G can be summarized:

- Introduction of packet data -Digital systems for voice + rate data traffic
- Internet access through General Packet Radio Systems (GPRS)
- Better modulation techniques with Enhanced Data Rates for Global Evolution (EDGE)

III.Third Generation (3G)

The 3G mobile systems used digital modulation but apart from voice which was a major thrust in 2G, focused mainly on high-speed data including compatibility with the internet. The 3G standard utilized Universal Mobile Telecommunications System (UMTS)/W-CDMA as well as CDMA 2000 as its core network architecture, By using packet switching, technology was improved to allow speeds up to a few Mbps whereas internet services browsing or checking emails became possible while driving around 60-80 km/hr speed. It also enabled multimedia transmission whereby downloading and watching a movie clip on a mobile phone became realistic. International Mobile Telecommunications-2000 (IMT-2000) were the specifications by the International Telecommunication Union for the 3G network. However, 3G technologies faced constraints in enabling the above-mentioned multimedia features in real time. The associated problems identified were regarding fast fading and Doppler effects. Some of the main features of 3G can be summarized as:

- Speeds 384 Kbps to 2 Mbps
- Improved spectral efficiency with Increased bandwidth and data transfer rates
- Multimedia transmission with limits i.e up to Sending/receiving large email messages
- Larger capacities compared to 2G and 2.5 G and broadband capabilities

IV. Fourth Generation (4G)

An improvement in 3G was its 3.5G variant named High-Speed Downlink Access (HSDA) which referred to the combination of High-speed Downlink Packet Access (HSDPA) and

High-speed Uplink Packet Access (HSUPA). HSDPA allows data rates up to 14.4 Mbit/s (Mbps) in the downlink. HSUPA makes uplink data rates of 5.76 Mbit/s possible. The main difference between 3G and 4G is the data rate due to changes in technology. The key technologies that made 4G possible were the introduction of Multiple Input Multiple Output (MIMO) and Orthogonal Frequency Division Multiplexing (OFDM). The most important 4G standards are Worldwide Interoperability for Microwave Access (WiMAX) and LTE. LTE was developed after WiMAX with similar technology. As a business module, major cellular operators favoured LTE and major world countries started issuing 4G licenses for 4G using developed LTE systems. However, the licensing cost was high for migrating from 3G to 4G and it was a shift from low data rates for the Internet to high-speed data rates for mobile video.

Even after it was widely available, many networks were not up to the required speed of 4G. 4G LTE is a "fourth generation long term evolution", capable of delivering a very fast and secure internet connection. 4G is the predetermined standard for mobile network connections. 4G LTE is the term given to the path which has to be followed to achieve those predefined standards. Some of the features of 4G LTE are:

- Support interactive multimedia, voice, video
- High speed, high capacity and low cost per bit (Speeds of up to 20 Mbps or more.)
- Global and scalable mobile networks.
- Ad hoc and multi-hop networks

V. Fifth Generation (5G)

In recent years, cellular network technology has passed through a paradigm shift in its deployment and also in the optimization of resources. 5G services aim to fuel machine-to-machine communications or Internet of Things (IoT) services in addition to people-to-people connectivity. In this way, it will connect people with various systems in smart environments. 5G networks can transport a very large amount of data extremely fast and are still reliable across a large number of devices. 5G technologies can support applications such as smart homes, smart buildings, smart cities, cloud-based services, remote robotics-based services, virtual reality, augmented reality, and services related to industry automation. New infrastructure elements like femto/pico base stations, electronics systems such as fixed/ mobile relays, cognitive radios and distributed antennas are undergoing tremendous deployment. These advancements are trying to make 5G cellular networks more heterogeneous. In some cases, such implementations may present challenges but still, these new functionalities and new services necessitate a new way of deploying advanced mobile services. A comparison of various generations of mobile wireless systems concerning various aspects is summarized in Table 1.1

Table 1.1 Comparison of 2G, 3G, 4G and 5G

Aspects	2G	3G	4G	5G
Introductory Year	1993	2001	2009	2018
Technology	GSM	WCDMA	LTE,WiMAX	MIMO, mm Waves
Access System	TDMA, CDMA	CDMA	CDMA	OFDM,BDMA
Switching Type	Circuit switching for voice and packet switching for data	Packet switching except for air interference	Packet switching	Packet switching
Internet Service	Narrowband	Broadband	Ultra broadband	Wireless World Wide Web
Bandwidth	25 MHz	25 MHz	100 MHz	lower bands (sub-6 GHz) + mm Wave

Aspects	2G	3G	4G	5G
				(30 GHz-300 GHz)
Advantage	Multimedia features like SMS & MMS, internet access and SIM introduced	High security, international roaming	Speed, high speed handoffs, global mobility	Extremely high speed and low latency
Application	Voice calls, short messages	Video conferencing, mobile TV, GPS	Mobile TV, wearable devices	High resolution video streaming, remote control of vehicles, robots and medical procedures
Data rate	9.6 /14.4 Kbps	500-700 Kbps	100-300 Mbps	20 Gbps peak data rates and 100+ Mbps avg.

1.4. Cellular concepts - fundamentals of system design

Early mobile radio systems prioritized wide coverage using solely, powerful transmitters, which limited the reuse of frequencies due to interference. To overcome this obstacle, the reorganization was necessary to maintain wide coverage while increasing capacity within the limited spectrum.

1.4.1. Introduction

The cellular concept revolutionized spectral congestion and user capacity by using multiple low-power transmitters (small cells) instead of a single high-power transmitter (large cell). To reduce interference, each base station has designated channels that span a small area. High capacity in sparsely allocated spectrum and effective channel distribution over the region is made possible by this approach.

To avoid greater interference, base station counts can be increased while transmitter power is decreased in a particular market as service demand increases and more channels become necessary. Without requiring additional radio spectrum, this tactic provides increased radio capacity. This fundamental idea of reusing same fixed frequency channels to serve a large number in a given service area is the foundation of all contemporary wireless communication systems. Moreover, the cellular concept ensures that every subscriber device within a nation or continent has the same channel set, enabling seamless regional operation of mobile devices.

1.4.2. Frequency reuse

To control coverage areas and minimize interference, cellular radio systems employ sophisticated channel allocation and reuse inside cells. To ensure that adjacent cells use separate channels, base stations are assigned specific channel groups. Frequency planning, also known as the frequency reuse scheme, maximizes channel allocation for effective communication inside the cellular network. Figure 1.10 illustrates cellular frequency reuse by showing cells that share a group of channels. In cellular systems, the hexagonal cell form is frequently utilized for radio coverage analysis. It offers a methodical layout for the coverage regions of base stations. The hexagon is selected because, in comparison to a square or equilateral triangle, it has the biggest area and can cover areas without overlap or gaps. The weaker mobiles at the cell's edge must be accommodated by the cell design. The hexagon form is similar to a circular radiation pattern and enables the least amount of cells to cover a given geographic area.

To understand the idea of frequency reuse, assume a cellular system that has S duplex channels available for use. The total number of radio channels (S) that are available can be defined as follows: Consider that each of given N cells is comprising k channels (where $k < S$). Further, consider that (S) channels are distributed among N cells with different groups of channels, each of N with an equal number of channels. Relation among,S, k and N is given by

$$S = kN \tag{1.1}$$

Each group of N Cells are known as a cluster and each cluster is utilizing the full frequency spectrum. When such cluster is duplicated M number of times can serve as a capacity metric (C) and is defined as:

$$C = MkN = MS \tag{1.2}$$

It should be noted that S is a constant number (number of duplex channels) and hence the capacity of a cellular system, as indicated by Equation (1.2), is directly linked to the frequency of cluster replication (M) within a fixed service area. The cluster size, denoted by N, is typically set at 4, 7, or 12. Decreasing N (number of cells in the cluster) while maintaining the cell size leads to a greater number of clusters needed to cover the same area, resulting in increased capacity (larger C). A larger cluster size signifies a greater ratio between the distance between co-channel (Co-H) cells (D) and cell radius(R), while a smaller cluster size indicates closer proximity of Co-H cells that may bring Co-Channel Interference (CCI) related problems. The choice of N depends on the interference tolerance of mobile or base stations while ensuring communication quality. To optimize capacity over a specific coverage area, it is preferable to minimize N.

The hexagonal geometry in Figure 1.10 features six equidistant neighbors and lines connecting the centers of each cell to its neighbors at intervals of 60 degrees. This restricts the possible cluster sizes and cell arrangements. In hexagonal geometry, to ensure seamless connections between neighboring cells, the number of cells per cluster, denoted as N, must follow Equation (1.3).

$$N = i^2 + ij + j^2 \tag{1.3}$$

Where i and j are non-negative integers. To move to the closest Co-H neighbors of a specific cell, the following steps are required:

First step: Perform cell center to center Movement for i cells along any adjacent hexagons.

Second step: Take 60 degrees counter-clockwise turn then and move j cells.

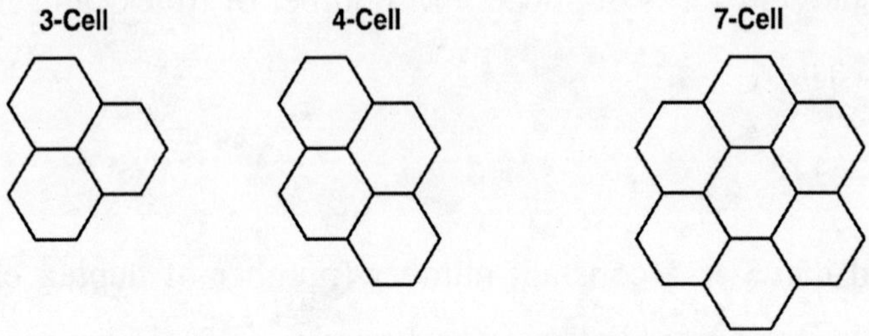

Figure 1.10 Conceptual diagram of Cellular Frequency Reuse.

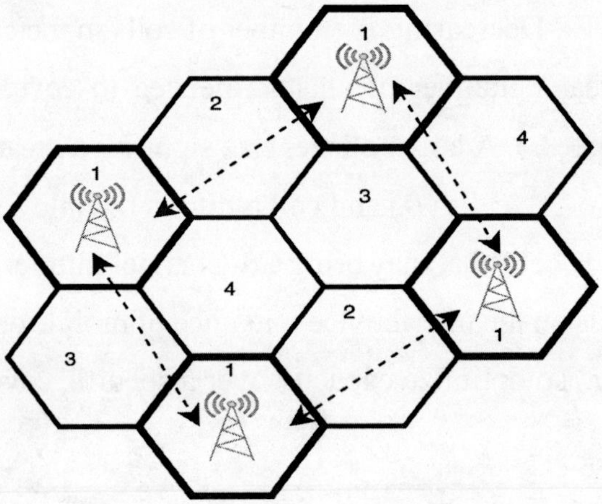

Figure 1.11 Co-channel (CCH) cell location found in a cellular system (reuse factor = 4)

Figure 1.11 illustrates this for i = 2 and j=0 (for instance, N=4). It should be noted that the number of cells per cluster takes only a few (not all) integer values. This is due to dependence on values of i and j (integer values) as shown in Equation 1.3. N takes some integer values like 3, 4, 7 , 9 , 12 but no other integer values like 5,6,8 etc.

1.4.3. Channel Assignment (CA) Types

To maximize radio spectrum utilization, a frequency reuse scheme must be put in place to reduce interference and increase capacity. System performance is greatly impacted by the choice of CA schemes. Strategies for CA can be either fixed or dynamic. Simpler is fixed CA only one-time planning is required. On the other hand, in dynamic assignment strategies, system planning has to be adaptive. The system's performance is impacted by the CA mechanism selected, particularly when handling calls between different cells.

Every cell in a fixed CA scheme is given a set number of voice channels in advance. Incoming calls are rejected when every channel is in use. Likewise, any request for call setup made from within the cell will only be taken into consideration if there are available channels. The call is blocked if the cell's available channels are all occupied. However, to prevent call interruptions, cells can use the borrowing method to borrow channels from nearby cells while being supervised by MSCs.

Voice channels are not permanently assigned to particular cells when using a dynamic channel assignment technique. In response to a call request, the serving BS requests the MSC for a channel. The channel is subsequently assigned by MSC to the call request taking into account various cost functions, namely frequency of channel usage, potential future call blocking within the cell, and other variables. The MSC assigns a frequency only if it's not currently in use in the cell or any nearby cells within the restricted frequency reuse zone to prevent interference. Employing dynamic CA helps reduce the chances of blockages, enhancing the system's trunking capacity by enabling access to all available channels in the market for all cells. The MSC must continuously collect channel occupancy-related data, radio signal strength and traffic distribution for every channel to properly execute dynamic channel assignment. Although this increases the system's storage and computational demands, it boosts channel utilization while reducing the likelihood of call blockages.

1.4.4. Types of Handoff

When an MS navigates into a different cell during a call, MSC initiates the process to transfer the call to a new channel that is part of the new base station. The handoff method, in theory, entails locating a new base station and allocating the necessary channels (voice and control) to the identified new base station. It's important to remember that handoffs must be seamless for the user and executed as infrequently as feasible. Handoffs are an essential task in cellular radio systems. Based on the variation of signal strength received from either base station, the decision should be taken about the optimum handoff interval. Handoff interval time (size of the region is important and dependent upon the velocity at which the user is travelling. If the user is travelling too fast, there will be little time to handoff and if it is not done properly, the call may get dropped during the handoff. When assigning vacant channels in a cell site, any cellular radio system must give priority to handoff requests over call initiation requests. The concept of handoff can be understood in Figure 1.12.

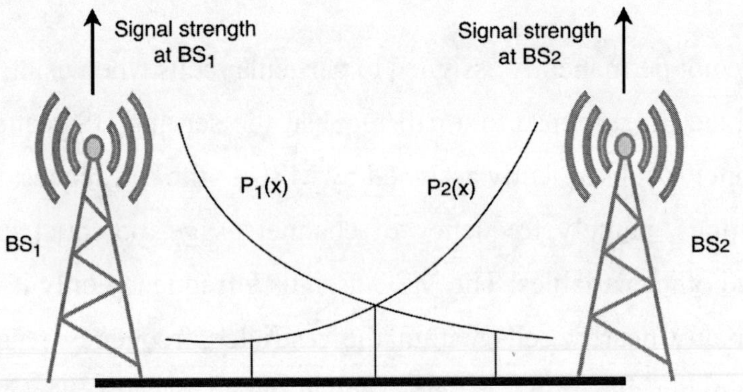

Figure 1. 12 Conceptual diagram showing need of Handoff mechanism

As shown in Figure 1.12, suppose we have base station *I* (shown on the left side) and we have another base station *J* (shown on the right side). Both base stations are located a certain distance apart. Let us consider a case when a mobile station is currently communicating with base station *I* and also travelling towards base station *J*. During the travel process, the signal strength from the base station *I* starts dropping and as it goes more and more towards base station *J*, it (signal strength from base station *I*) goes towards an acceptable region where minimum sensitivity of MS threshold exists for call sustainment. The ideal signal level at which to start a handoff must be specified by system designers to comply with these requirements. Before deciding whether to handoff, it is important to be sure that the mobile device is indeed going out of service range of the presently serving BS and that the reported signal level reduction is not the product of temporary fading. The base station optimizes the running average measurement of signal strength to prevent needless handoffs and finish necessary handoffs before a call is terminated owing to low signal strength. Before starting a handoff, BS observes the signal strength for a certain time interval.

How long a particular user remains in the network depends on several time-varying factors, including the propagation constant of a given area, interference faced by MS from nearby cells, the distance between the subscriber and BS etc. In first-generation analogue cellular systems, the MSC supervises the base stations' measurement of the signal strength. Mobile devices help handoff decisions in digital TDMA systems of the second generation. Every mobile station participating in Mobile Assisted Handoff (MAHO) regularly transmits the results to the serving base station by measuring the power received from nearby base stations. A handoff occurs when the power level received from the BS of a nearby cell exceeds the power received from the current BS in two parameters namely amount plus period.

By setting aside certain channels for handoff requests, guard channels minimize overall traffic while optimizing spectrum efficiency through dynamic channel assignment. This is one way to prioritize handoffs. Another strategy is to balance the risk of termination and overall traffic by waiting for handoff requests to lower the likelihood of call termination. The duration between a signal level drop and a call termination is the basis for queuing, and traffic patterns dictate the size of the queue. Queuing does not, however, remove the possibility of a call being terminated abruptly because lengthy delays still exist.

In contrast to channelized wireless systems, which switch channels during a handoff (sometimes called a hard handoff), spread spectrum mobiles use the same channel in each cell. Consequently, a handoff is not a channel shift; rather, it relates to the management of the radio transmission by a separate BS. The most effective signal version may be found by the MSC by analyzing signals from a single subscriber across several adjacent base stations in real-time. By making use of the spatial variety resulting from the different BS locations, this technique allows the MSC to "soft handoff" which refers to the method of selecting signals from various base stations.

1.4.5. Interference and System Capacity

Interference is one of the primary concerns in cellular communication, and it can be produced by a variety of sources including neighboring mobile devices, base stations, and non-cellular systems. Interference can cause issues in voice channels as well as CCHs. These include crosstalk and noise in the background on voice channels whereas, missed calls, black calls or drops of ongoing calls in CCHs.

Urban environments are particularly prone to interference because of high levels of RF noise plus the density of mobile and base station sites. It is difficult to regulate interference from both inside and outside the system, particularly from transmitters belonging to competing cellular carriers that are located close together. Common examples of interference namely Co-channel Interference (CCI) and Adjacent Channel Interference (ACI) will be dealt with in the next subsections.

1.4.5.1. Co-channel interference (CCI)

Frequency reuse enables numerous cells in a coverage zone to share the same frequencies that may result in Co-H cells and CCI. Unlike thermal noise, boosting carrier power cannot overcome CCI because it causes greater interference with surrounding cells. Physical separation between Co-H cells is required to reduce interference. The conceptual diagram of the Co-H cells is shown in Figure 1.13.

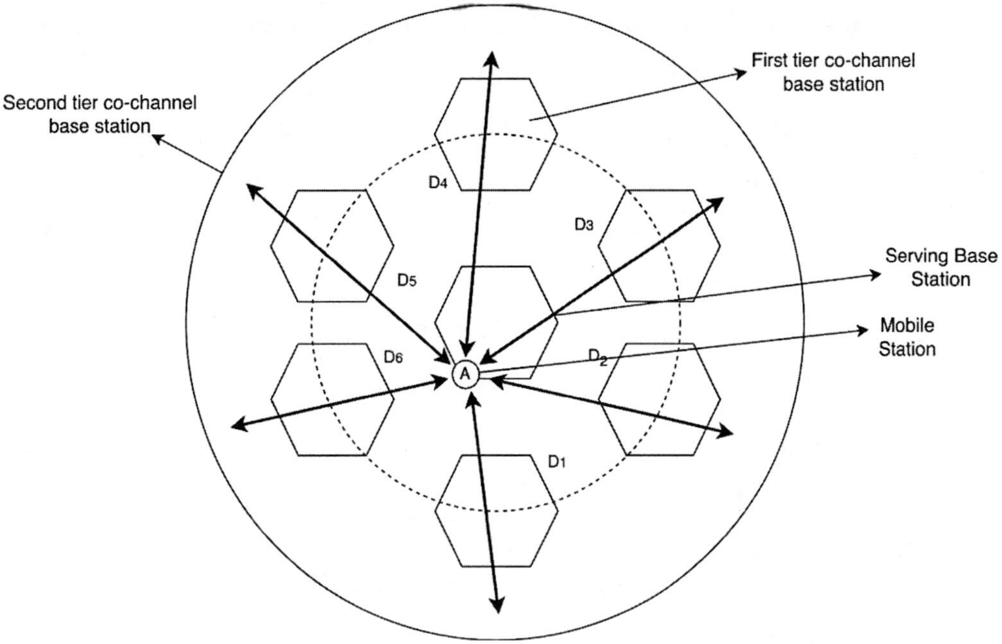

Figure 1.13 Conceptual diagram showing Co-channel (Co-H) cells

An example of CCI occurrence is depicted in Figure. 1.13. It can be seen that in the centre, a hexagonal cell with its base station is located. For theoretical simplicity, hexagonal cells are shown however, in real life cells will be irregular in shape. To optimize limited bandwidth, frequency bands are reused; there will be several Co-H cells. One cannot put a MS first inside the cell in the center.

Using the frequency band allocated, MS is free to move around within the cell. The first Co-Hl cell is located at a distance from the original cell. There is not just one Co-H cell but there should be several Co-H cells of radius R. It is seen that 6 CCI cells exist in the first tier.

It also can be noted that the location of the MS will determine the exact distance of the mobile from the interfering base station. Six BSs radiating in the same frequency band are shown as D1, D2…,D6. They all are located at different places and considered to be Co-H BS within the first tier but please remember the hexagonal cell pattern is being repeated throughout the area of interest. There will be a second tier as well but it is clear that as the distance increases the resulting CCI levels will also go down. Sometimes, second-tier CCI (though will be lesser compared to tier I) also may have to be considered, if required.

It further can be noted that, unlike thermal noise, CCI cannot be perfectly mitigated by increasing the power of a carrier signal that is sent to the transmitter. In other words, in order to improve S to I (the signal to interference) ratio, merely increasing the strength of a desired signal may not work as a signals (for a given user at a particular frequency) in one cell will be considered as interference in its Co-H cell (other user) and that degrades the performance

Let us consider that each cell is of similar size (and shape) and each BS transmitter (located in the center of a cell) transmits an equal amount of power.

The CCI ratio is defined by the distance between the nearest Co-H cell centres (D) to the cell radius (R). The spatial space between Co-Hl cells grows in proportion to the cell coverage distance as the D/R ratio increases. This reduces interference by better isolating RF energy from neighboring Co-H cells.

The co-channel reuse ratio, given as parameter Q, is related to cluster size (N). For a hexagonal geometry, this ratio is given by

$$Q = D/R = \sqrt{(3N)} \tag{1.4}$$

A small value of Q results in greater capacity when N is small. Conversely, larger N results in a large value of Q enhances transmission quality by reducing CCI. Balancing these two objectives is crucial in practical cellular design. Table 1.2: shows Co-channel Reuse Ratio (Q) for Some Values of N (size of a cluster).

Table 1. 2 Co-channel Reuse Ratio for Some Values of N

	Cluster Size (N)	**Co-channel Reuse Ratio (Q)**
$i=1,\ j=1$	3	3
$i=1,\ j=2$	7	4.58
$i=2,\ j=2$	12	6
$i=1,\ j=3$	13	6.24

Signal-to-interference ratio (S/I or SIR) for a mobile receiver monitoring a forward channel can be defined as:

$$\frac{S}{I} = \frac{S}{\sum Ii} \tag{1.5}$$

Where,

S represents the desired signal strength (power) from the target BS to given MS. I_i indicates the interference signal strength (power) generated by BS of ith interfering Co-H cell. Once signal strengths of Co-H cells are identifiable, the S/I ratio for the forward link (BS-> MS) can be calculated using Equation (1.5).

The average received signal intensity in a mobile radio channel declines with distance between transmitter and receiver according to a power law, according to propagation measurements. At a distance d from the transmitting antenna, the estimated average received power P_r can be roughly calculated as follows:

$$Pr = Po * (\frac{d}{do})^{-n} \tag{1.6}$$

Or

$$\Pr(dBm) = Po(dBm) - 10 * n * \log\left(\frac{d}{do}\right) \tag{1.7}$$

Where,

- Po denotes power signal value received at a nearby reference point
- $d0$ is reference distance from Tx antenna,
- n refers to path loss exponent.

In the forward link situation, Co-H BS cause interference, whereas the desired signal is coming from the serving BS. The received power at a particular MS from the ith interfering Co-H cell is proportional to $(Di)^{-n}$ where, Di represents the distance between the ith interference and MS. In urban cellular systems, the route loss exponent usually lies in the range of 2 to 5.

When each BS's transmitted power is equal and the path loss exponent is the same throughout the coverage area, S/I for a mobile can be defined as:

$$S/I = R^{-n}/\sqrt{(Di)}^{\,-n} \tag{1.8}$$

Equation 1.8 can be simplified when just taking into account the first layer of six interfering cells and assumes that all interfering BSs are equally spaced distance D from the target BS, we can rewrite earlier equation as :

$$S/I = (\sqrt{3N})^{n/6} \quad \text{Or} \quad S/I = (D/R)^{n/6} \tag{1.9}$$

The relationship between the cluster size N and the system's overall capacity is described by Equation .1.9, which is derived with the help of Equation .1.2. Consider a scenario where the six nearest cells are close enough to cause significant interference and are equidistant from the base station. If a cellular system is utilizing 30 kHz FM channels, desired voice quality threshold is met when S/I is 18 dB or higher. By applying Equation 1.9, it is evident that a minimum cluster size of 7, with $N \geq 6.49$, is necessary to meet this criterion, assuming a path loss exponent of $n = 4$. It should be mentioned that Equation.1.9 is predicated on the assumption that all interfering cells are equally spaced from the base station, which could produce overly optimistic outcomes in some circumstances. In frequency reuse plans such as $N = 4$, the nearest interfering cells may vary significantly in distance from the target cell.

Figure 1.14 indicates that in a 7-cell cluster, when the MS is at the border region of a given cell. The mobile device is located at a distance of D-R from the two closest CCI cells, approximately at distances of $D + R$ from the farthest two cells and still exactly $2D$ distance from the rest two interfering cells in the first tier. Utilizing these detailed knowledge and assuming n is equal to 4, the signal-to-interference ratio can be closely estimated using equation (1.9) as:

$$\frac{S}{I} = \frac{1}{2(Q-1)^{-4} + 2(Q+1)^{-4} + 2Q^{-4}} \tag{1.10}$$

Equation 1.10 can be written again in terms of the Q, as:

$$\frac{S}{I} = \frac{R^{-4}}{2(D-R)^{-4} + 2(D+R)^{-4} + 2D^{-4}} \tag{1.11}$$

To understand the importance of Co-H interference let us analyze case for $N = 7$. It can be easily computed from Equation 1.9 that the co-channel reuse ratio is 4.6. For the worst-case, S/I is approximately 49.56 (17 dB) based on Equation 1.11. However, if mobile specifications suggest that for a 7-cell cluster, the S/I ratio should be at least 18 dB in the worst-case scenario. In such case, To ensure proper performance in the worst-case scenario, increasing N to the next largest size, which is determined to be 21 (corresponding to $i = j = 2$) from Equation 1.3, becomes necessary.

It is important to note that this adjustment, however, results in a capacity reduction since a 12-cell reuse provides a spectrum utilization of 1/12 per cell, whereas a 7-cell reuse offers 1/7. In reality, a capacity reduction of 7/12 would not be acceptable to cater for the seldom-occurring worst-case scenario. The discussion above highlights that CCI plays a crucial role in determining link performance, influencing the frequency reuse strategy and the capacity of given systems. The same fact can be endorsed by the following example.

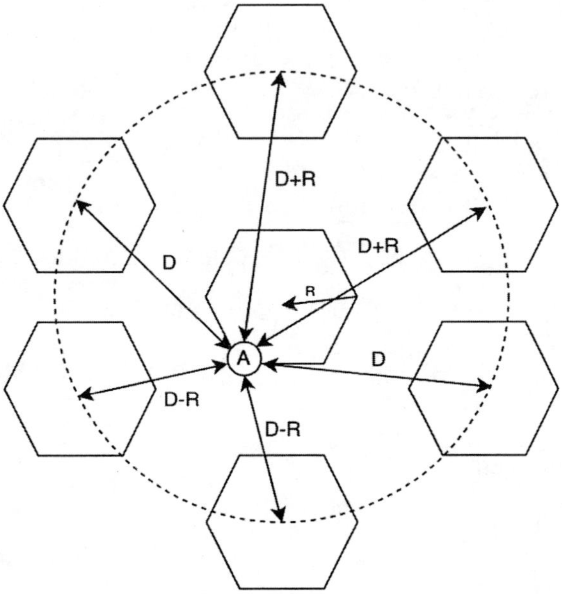

Figure 1.14 cellular size N=7, the first tier of co-channel cells

In Figure 1.14 for cellular size N=7, the first tier of co-channel cells is displayed. The MS experiences the worst-case Co-H interference when it reaches the cell boundary (point A).

Example 1.1: In a given cellular mobile system, in order that mobile receiver works properly, a signal-to-interference ratio should be 15 dB or better (i.e. higher). The system is deployed in two different scenarios where the path loss exponents are (a) zeta = 4 and (b) zeta = 3, respectively. It is assumed that (i) interference is coming from the first tier; only (ii) six cells will be in the first tier with the same distance from MS. Check the suitability of cluster size equal to 7 for satisfactory operation of the system.

Solution: Given data :

signal-to-interference ratio should be 15 dB or better

path loss exponents are (a) zeta = 4 (b) zeta = 3, respectively

Interference is coming from first tier only and there are six CCH cells in the first tier a) Path loss exponent zeta = 4.

Let us first consider the suitability of the 7-cell reuse pattern.

co-channel reuse ratio

D/R=4.583. (By using Equation 1.4)

signal-to-noise interference ratio

S/I= 75.3=18.66dB (By using Equation 1.9)

Here, the computed signal-to-noise interference ratio is greater (better) than the minimum required S/I, N=7 by the receiver circuit, 7-cell reuse pattern can be used. It is left to the reader that if one computes with 3 or 4 reuse distance computed signal-to-noise interference ratio will be lower (worse) than the minimum required S/I by receiver circuits and hence that cannot be used.

b) Path loss exponent zeta = 3

Again, let us start with standard 7-cell reuse pattern,

co-channel reuse ratio

D/R=4.583. (By Using Equation 1.4)

signal-to-noise interference ratio

S/I=16.04=12.05dB

Here, the computed signal-to-noise interference ratio is lesser (worse) than the minimum required by the receiver circuit, 7-cell reuse pattern cannot be used.

1.4.5.2. Adjacent channel interference (ACI)

Adjacent channel (ACH) interference occurs when nearby frequencies leak into the passband due to imperfect receiver filters. The near-far effect happens when a nearby transmitter interferes with a receiver's attempt to connect to a base station. This interference can disrupt cellular systems and impair the signal between mobile users and base stations. It can be decreased through proper filtering and strategic channel allocation.

Furthermore, MSs are assigned non-adjacent channels to reduce interference. Interference can be reduced greatly by increasing frequency separation inside a cell and avoiding contiguous frequency bands. Sequentially assigning channels across cells and avoiding adjacent channels in surrounding sites significantly reduces interference. One of the effects of ACI is called the near far effect. This happens when an interferer is close to the base station and also radiates in the adjacent channel. At the same time, actual subscriber is far away from the base station. In practice, the path loss exponent is close to four. In such a case, the signal strength goes down with the power of four of the distance. Here, subscriber signals will get a lot of interference at the base station. This phenomenon can be understood by Figure 1.15.

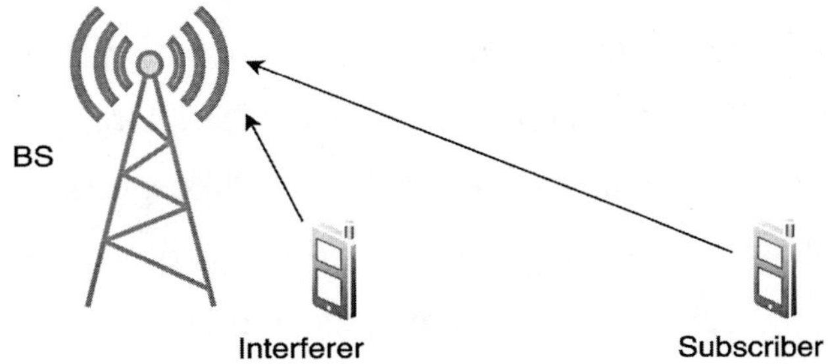

Figure 1. 15 Conceptual diagrams to illustrate Near-Far problem

As shown in Figure 1.15 considers a Base Station (BS) and the subscriber (S) that is located at a far distance from BS. System is having high traffic and the interfering handset is located much closer to the base station. Due to high traffic, the system has to allocate an adjacent frequency band. Suppose that the subscriber is trying to communicate with the base station but due to far distance (and path loss exponent being high) weak signal reaches the base station. Now we present a few strategies utilized either to reduce interference to increase capacity or to achieve both. The first technique is to have effective power control to reduce interference and keep it under control. In practical mobile systems, the power level of each subscriber handset is under constant monitoring and control by the serving BS. This is applicable when MS is approaching the BS going away. The power control mechanism serves not only reduction in interference level but also to prolong battery life. In later chapters, we shall show that in CDMA techniques power control is a key feature to increase the capacity of the system. It is worth noting that with an increase in demand for the system, service operators have to plan for more channels per unit coverage area. Some of the common techniques of improving capacity are cell splitting, sectioning and microcell zoning which are discussed next.

1.4.5.3. Cell sectorization

Cell sectoring focuses on reducing the D-to-R ratio without changing cell size, which affects interference levels. By decreasing the number of cells per cluster, frequency reuse increases but also leads to higher interference. Sectoring aims to selectively reduce reuse distance to boost capacity while managing interference. The strategy involves reducing relative interference without altering transmitted power to improve system performance.

Cell sectoring is a method of reducing Co-channel interference by replacing a single omnidirectional antenna with several directional antennas each radiating within a specified sector. This approach reduces interference by 1/6th, as directional antennas transmit and receive only a fraction of the total number of Co-H services. A cell is typically partitioned into sectors, which can be 120 degrees, 4-90 degrees, or 6-60 degrees. As the number of sectors increases, the amount of Co-channel interference picked up decreases. However, this comes at a stronger price, as more sectored antennas are needed. Additionally, each time a mobile station moves from one sector to another sector, there must be a hand-off, which comes at its own cost. To minimize interference, it is essential to limit the number of mobile stations handing off at a time.

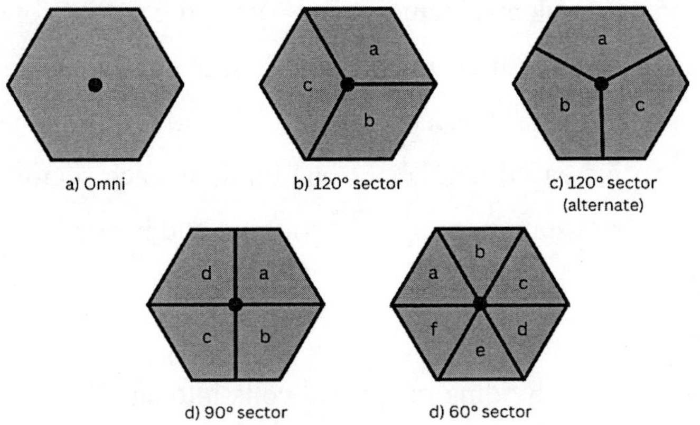

Figure 1.16 Cell sectioning of various geometries.

Cell sectioning can be achieved using various geometries, such as omnidirectional, 90 degree, 120 degree, or 60 degree sectors, which can be seen in Figure 1.16.

The standard omnidirectional geometry allows one base station to radiate at all portions of the cell. However, subdividing the hexagon into three sectors can create slight overlap between the sectors, allowing for successful hand-offs. In this case, a 120 degree sector was used. Other methods include subdividing the cell in another way, requiring an average distance from the serving base station antenna. 90-degree sector antennas are also used, as seen in IEEE 802.16, IEEE 802.16, and Local multipoint distribution service (LMDS), local to multi-point distribution services. A 60 degree sector is another option, dividing the original cell into six sectors of 60 degrees each. However, anything below this is not preferred as too many sectors would require too many hand-offs and antennas. In practice, these methodologies are used to ensure successful hand-offs and maintain a clear distinction between a, b, and c sectors.

Sectoring involves placing directional transmitters at corners where adjacent cells meet, increasing capacity. This approach differs from cell splitting, as the base station is placed at a corner. The new sector base station supports 1,2, and 3, and can support all remaining corner points. The reuse distance decreases, and the coverage area increases. Sectoring improves the signal to interference ratio by cutting off interference from five other Co-H cells, resulting in a 120-degree area of coverage.

Cell sectioning has several problems, including increasing the number of antennas at each base station, which is expensive, requires maintenance, and consumes power. This decreases trunking efficiency, as each sector uses a portion of the pool, resulting in decreased efficiency. Additionally, there is an increased number of handoffs, as each sector is treated separately. However, many modern base stations support sectioning and handoffs independently, making it an alternative to cell splitting.

1.4.5.4. Cell splitting

Cell splitting is the process of dividing congested cells into smaller ones, allowing for better service and reduced bandwidth, which is illustrated in Figure 1.17. In situations where a cell is too congested to provide enough service, cell splitting can be used to reduce the number of base stations and transmit power. This reduces the size of the cell, requiring more cells to be used, leading to more clusters and channels. This allows for a system capacity increase by replacing large cells with smaller ones without disturbing channel allocation. However, frequency planning requires significant effort and resource allocation. By dividing cells into smaller cells, the system can grow without upsetting channel allocation, ensuring better service and reduced bandwidth.

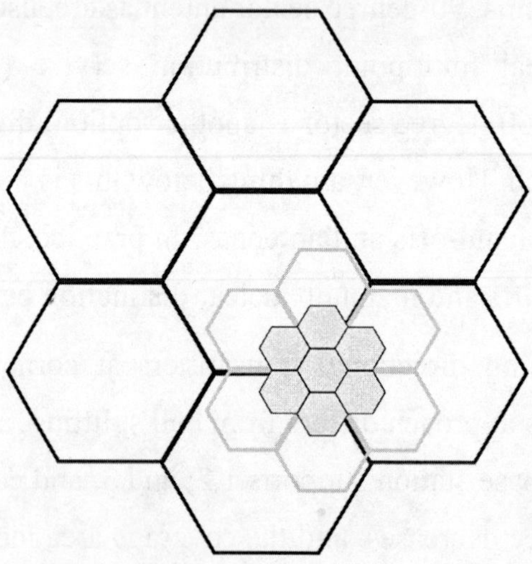

Figure 1.17 Illustration of cell splitting.

For example, large cells were initially used in low-density areas due to low demand. However, as users increased, some regions needed more cells. To accommodate this, cells were partially split into smaller cells, which could handle higher density.

If more users were needed per cell, smaller base stations were added to increase capacity. The frequency was reused without additional bandwidth. Smaller cells could be activated or deactivated based on traffic patterns, allowing for efficient resource use.

The choice was to continue with smaller cells or deactivate them if demand decreased. Smaller cells could be activated during peak office hours and deactivated when demand decreases. The question is not whether cell sizes are dynamic, but rather the boundaries of the cells. The goal is not to manipulate cell size or boundaries.

Unit Summary

In this unit, a module on the Evolution of Wireless Communication provides a comprehensive overview, beginning with the historical roots of wireless communication and progressing through the advancements from 2G to 5G technologies.

It explores mobile radio systems, including Paging, Cordless Telephony, and Cellular Mobile Systems, offering practical examples for real-world application.

A crucial aspect of the module focuses on the fundamentals of system design, covering Frequency Reuse, Channel Assignment, Handoff types, and the management of Interference and System Capacity.

The discussion emphasizes the importance of these principles in designing efficient and robust wireless communication systems.

The chapter further delves into key concepts, such as Handoffs, Frequency Reuse and the system's channel planning.

Handoffs play a vital role in seamless mobile traffic transition between cells, and the module explores various implementation methods.

The system's capacity, influenced by S/I limits, determines the frequency reuse factor, impacting the number of channels within a coverage area.

Additionally, the module presents techniques to increase capacity by increasing S/I, such as sectioning and cell splitting.

Increasing the system's user base is the goal shared by all of these strategies. In turn, the features of radio propagation in real-world system scenarios affect how effective these methods are.

Exercises

Multiple Choice Questions

1) What is the bandwidth of the 4G network?

a) 25 MHz b) 100 MHz c) 100 GHz d) 300 GHz

2) What is the role of the Mobile Switching Center (MSC) in a cellular telephone system?

a) The MSC connects base stations to the Public Switched Telephone Network (PSTN).

b) The MSC coordinates base station activities and handles system maintenance.

c) The MSC supports full duplex communications between mobile users.

d) The MSC facilitates the handoff process during calls.

3) What is the key objective of 5G cellular networks?

a) Enhancing traditional voice and text communication services.

b) Focusing solely on people-to-people connectivity.

c) Fueling machine-to-machine communications and Internet of Things (IoT) services.

d) Limiting the amount of data transported to ensure reliability.

4) What is a key advantage of dynamic channel assignment over fixed channel assignment in cellular networks?

a) Dynamic channel assignment enhances trunking capacity and reduces call blockages.

b) Fixed channel assignment ensures a higher level of system performance.

c) Dynamic channel assignment reduces the chances of call handoffs between cells.

d) Fixed channel assignment allows for borrowing channels from nearby cells during congestion.

5) What is the primary purpose of cell sectioning in cellular networks?

  a) To increase cell size and coverage area.

  b) To decrease the number of sectors per cell.

  c) To minimize the need for handoffs between cells.

  d) To reduce interference levels and boost capacity.

6) Given a cellular system with N cells and k channels allotted to each cell, what is the total number of radio channels (S) that are available?

  a) S=k*N   b) S=k/N   c) S=N/k   d) S=k^N

7) In any cellular system the frequency reuse factor for N cells ?

  a) N   b) N^2   c) 2*N   d) 1/N

8) Capacity of any cellular system is directly proportional to __________

  a) Cells in system   b) Replication of clusters

  c) Base Stations employed   d) Users

9) How many channels are available per cell with a 4-cell reuse factor in a cellular system that uses two 25 KHz simplex channels for each full-duplex voice channel ad has a 30 MHz spectrum allocation?

a) 150   b) 600   c) 50   d) 85

10) State the condition of any handoff.

a) Movement of the users in various cells while on call

b) User in the same cell throughout the conversation

c) Movement of user in various cells in spite of being idle

d) User movement restricted to the same cell being idle

Answers to Multiple Choice Questions

1) b 2) b 3) c 4) a 5) d 6) a 7) d 8)b 9) a 10) a

Short and Long Answer Type Questions

Short Questions

1. Define the following terminologies in view of a cellular wireless communications system.

(a) Half duplex system (b) Handoff (c) Mobile switching centre (d) Control channel (e) traffic channel (f) frequency division duplex (g) cell

2. Explain paging systems with a suitable diagram.
3. Explain in brief, reasons for the selection of 'hexagonal structure' compared to other regular structures (like circles, squares /rectangles or triangles) for cell coverage.
4. Compare the following multiple access techniques conceptually.
   (a) TDMA Vs. FDMA
   (b) CDMA Vs. TDMA /FDMA
5. Briefly explain the following generation of mobile cellular networks.
   (a) 2G (b) 3G (c) 4G (d) 5G
6. Define Handoff. Explain the different types of handoffs.
7. Explain the concept of adjacent channel interference.

Long Questions

1. Explain mobile radio systems with examples and suitable diagrams.
2. In terms of various aspects, compare 2G, 3G, 4G and 5G of mobile communication.
3. Explain in detail how a cellular call is made.
4. Explain with the help of conceptual diagrams, all basic types of multiple access schemes.
5. Define frequency reuse. Explain the concept of frequency reuse with the help of a suitable diagram. List down the required equation showing the relationship between frequency reuse distance (D) in terms of the radius (R) of a cell and frequency resume factor (K). Show how the frequency reuse factor can improve the capacity of a cellular system.

6. Explain co-channel interference (CCI) in cellular mobile systems. Derive the relation between (S/I) in terms of number of cells in a cluster (N). Assume interference coming from (a) the first tier and (b) the first two tiers.
7. Explain the following concepts to improve the performance of cellular wireless communication systems. (a) Cell Sectorization (b) Cell splitting

NUMERICAL PROBLEMS

(1) Compute frequency reuse factor (N) and cluster size that should be used for maximum capacity for the following cases of signal to interference ratio: (a) 5 dB (b) 10 dB and (c) 15 dB. Consider the path loss exponent is 3.

(2) Transmitters produce 100 W. Compute the transmitted power in following units : (a) dBm and (b) dBw. If 100 W is applied to an antenna (gain unity) with a 900 MHz carrier frequency, find the received power in dBm at a free space distance for 100m, 200m, 500m and 1 km. Also calculate received power for 10km, and 5km.

(3) Calculate above problem 1 for 4, 6, and 7 CCH cells in the first tier and all are the same distance from the mobile.

PRACTICAL

1. Write a program to calculate the Frequency Re-use Ratio (FRR) using user-specified inputs: Co-channel (Co-H) distance and radius of cell.
2. Given a frequency reuse factor, Write a program to calculate the Signal-to-Interference Ratio for the worst case i.e. MS is located at the cell boundary.

KNOW MORE

Evolution of Wireless Communication as a comprehensive overview, depicts historical roots about advancements from 2G to 5G technologies across the globe. Exploration about mobile radio systems from Pagers to Cordless Telephone to cellular Mobile Systems, helps developers for better know how practical examples for real-world application. These days engineers trying to make career in wireless domain must understand fundamentals of system design, covering Frequency Reuse, Channel Assignment, Handoff types, and the management of Interference and System Capacity.

Advancements in technology especially simulations has help new designers about technical knowhow of these principles in designing efficient and understanding robust wireless communication systems. Knowledge about key concepts, such as Handoffs, Frequency Reuse and the system's channel planning is vital even though such practical implementations has gone ages ahead in recent times. The system's capacity, influenced by S/I limits, determines the frequency reuse factor, impacting the number of channels within a coverage area. Additionally, the module presents techniques to increase capacity by increasing S/I, such as sectioning and cell splitting. Increasing the system's user base is the goal shared by all of these strategies. In turn, the features of radio propagation in real-world system scenarios affect how effective these methods are.

Dynamic QR Code for Further Reading

REFERENCES AND SUGGESTED READINGS

1. Theodore Rappaport - Wireless Communications, Principles and Practice-ISBN 0130422320 (Edition) PHI)
2. NPTEL course: https://archive.nptel.ac.in/courses/117/102/117102062/

2 Mobile Wireless Propagation

UNIT SPECIFICS

In this unit, following aspects are discussed:

- *Wireless Propagation including three basic propagation mechanisms*
- *Radio Propagation models and Link budget*
- *Concept of Noise figure and Free-space path loss*
- *Multipath Shadowing*

RATIONALE

This module provides in-depth knowledge of wireless propagation, covering many areas necessary to understand and enhance wireless communication systems. It covers topics such as link budget, which accounts for transmitter power, antenna gains, and losses when calculating signal strength.

The module also examines free-space route loss, which explains why signals in open areas deteriorate over extended distances. It discusses multipath fading, shadowing, fading margin, and shadowing margin of the receiver, providing tips and roadblocks to lessen signal deterioration. Through this in-depth research, students gain a solid understanding of wireless propagation phenomena, which is essential for designing and operating dependable wireless communication networks.

PRE-REQUISITES

Mathematics (Class XII) , Basics of Signal and Wave , Basics of Communication system

UNIT OUTCOMES

After going through this unit, students will be able to:

U2-O1: Describe wireless propagation concepts, including the mechanisms and factors that influence signal propagation in varied settings.

U2-O2: Evaluate wireless communication systems by applying link Budget.

U2-O3: Analyze noise figure and Free-space Path Loss to assess and optimize system performance, detecting sources of noise and signal attenuation.

U2-O4: Explain the impact of shadowing on wireless communication systems.

Unit-2 Outcomes	***MAPPING of UNIT OUTCOMES WITH COURSE OUTCOMES (1- Weak Correlation; 2- Medium correlation; 3- Strong Correlation)***				
	CO-1	***CO-2***	***CO-3***	***CO-4***	***CO-5***
U2-O1	2	1	1	1	1
U2-O2	1	2	3	2	--
U2-O3	1	1	2	3	--
U2-O4	--	2	2	1	2

2.1. Highlights of Mobile Wireless Propagations

Mobile radio channels pose significant limitations to wireless communication systems due to their hostile environment and uncertainties such as multipath propagation, attenuation, and scattering. The widest transmission path can be either direct Line of Sight (LOS) or Non-Line of Sight (NLOS) which has been obstructed by buildings, foliage, hills, or cars. The behavior of radio channels is generally random and often time-varying, making it crucial to model them effectively and it is to note that radio channels behave differently in different frequencies. Modelling radio channels is a difficult part of mobile radio designs, as design statistics are carried out based on fundamental measurements and these statistics change over time. Random channel models are suitable for lower frequency bands (900 MHz, 1000 MHz, 2.4 GHz), while deterministic channel models are suitable for higher frequencies (above 10 GHz). Hence, understanding propagation basics is essential for successful mobile radio design.

Electrons generate electromagnetic waves that travel through space, and an antenna connected to an electrical circuit efficiently broadcasts and receivers receive these waves from a distance. The size of the antenna is determined by its wavelength, with larger antennas being more effective in lower frequency bands. As frequency increases, antenna size reduces, enabling smaller mobile devices and antennas. Radio bands, microwave bands, infrared bands, and visible light bands are all examples of frequency spectrums that can be utilized for wireless communication. Lower frequency radio, on the other hand, can propagate even when there is no clear LOS. It should be noted that Information over channels can be transmitted by altering waveform attributes such as amplitude, frequency, and phase. Pulse Position Modulation (PPM) is used in ultra-wideband to transmit narrow pulses with a reference.

Radio waves are simple to create and they can be utilized for both indoor plus outdoor communications, and they can be tuned to higher frequencies, allowing satellites to communicate with base stations. However, radio waves vary in frequency, with higher frequencies operating more like light and having difficulties penetrating obstructions. They can be absorbed by raindrops, fog, and dust particles, impacting their spread. On the other hand, at low-frequency bands, they can travel through obstacles and experience power declines as the movement of BS is away from the source. The power loss is determined by the route loss exponent, which is proportional to building density and other factors.

Mobile radio propagation uses a variety of frequency bands including Very Low Frequency (VLF), Low Frequency (LF), and Medium Frequency (MF) bands that follow the ground. Radio waves are transmitted as signals from one place to another, with the Earth's surface serving as a guide. However, at High-Frequency (HF) bands, the Earth absorbs ground waves, resulting in ionospheric reflections. The Earth's ionosphere, which contains charged particles, can be reflected into numerous bands, causing multi-path propagation. Reflections are through multi-points and they produce multipath signals that can be either constructive or harmful. The issue with HF bands is that when sent, they are absorbed by the Earth, causing the receiver to divert the signal upwards for reflection and return. This fading is caused by multipath propagation, which allows signals to reach the receiver via both direct channels and reflections. Faded signals are produced when signals are absorbed by the Earth; fading will be examined in subsequent sections.

Defense applications continue to use HF bands, particularly VHF communication bands. A direct LOS path is desired for VHF communications since it allows for direct transmission without bending across the HF channel. However, reflected waves can produce multipath channels unless an antenna has a limited half-power beam width. Directional antennas are used to prevent interference and increase signal strength. An omnidirectional antenna creates current, which can cause fading and random fluctuations in signal strength. The directionality of an antenna, antenna gain, and Half-Power Beam Width (HPBW) are all related, and low gain means wasted power. They can also be used to reduce interference, however they present complications. Because of the kilometers' of separation, aligning a narrow half-power beam width can be problematic, and the sharp cutoff can interfere with the original signals. High-velocity winds can also force antennas off-center, leaving them vulnerable to damage. In VHF communications, waves tend to follow more direct paths, and LOS communications are necessary. Reflected waves may interfere with original (desired) signals and can corrupt them, but recovery can be achieved. Overall, directional antennas have both pros and cons, but they are essential for VHF communications due to their direct nature and the need for line of sight.

Radio channel modelling is often performed statistically to replicate the intended system, which is accomplished by utilizing measurement data, such as power meters in automobiles or indoor settings.

Various measurements are carried out by changing the distance from the transmitter BS. Ultra-Wideband (UBW) technology is also utilized in measurement campaigns to collect data based on the desired communication system and spectrum.

Data can be collected either in time by transmitting CW and measuring amplitude or in frequency domains, allowing for a sweep across the whole frequency range. Furthermore, deterministic modelling is more useful at higher frequencies, greater than 10 GHz, because there are no good statistical models to explain channel behavior. Researchers use various techniques to ascertain the actual received signal. Changes in antenna angles (tilts) can also be made to increase coverage area.

These all efforts are focused on maximizing transmission distance and increasing the reliability of signal transmission, so it is crucial to consider both parameters.

2.1.1. Free Space Propagation Model

When there exists a LOS path between the T-R, the received signal strength can be estimated using the free space propagation model. Free space propagation is commonly observed in satellite communication systems and microwave LOS radio links. The free space model estimates received power with the different distances between T-R. The Friis free space equation determines the received power by a reception antenna at a distance d from a transmitting antenna. This equation is defined as follows:

$$P_r(d) = \frac{P_t G_t G_r \lambda^2}{(4\pi)^2 d^2 L} \tag{2.1}$$

Where, P_t represents transmitted power $P_r(d)$ is received a power that is dependent on the Transmitter-Receiver (T-R) separation parameter (d) , G_t stands for the transmitter antenna gain, G_r for the receiver antenna gain, d is the T-R separation distance in meters, L represents the system loss factor $(L \geq 1)$, and λ denotes the wavelength in meters.

Equation 2.1 reveals that received power depends on many parameters rather than just the power of the transmitter P_t. Two gain factors namely antenna gain at the transmitter G_t and antenna gain at the receiver G_r affect received power. Using a directional antenna (compared to using an omnidirectional one) these gains will automatically go up. Directional antennas at BS can help to reduce interference and provide higher gain. The received power decreases with higher frequency bands, as the wavelength λ decreases. As the frequency increases, the size of the cell decreases for the same transmit power. The ultra-wideband (3.1-10.6 GHz) is intended to work up to ten meters. It should be noted that in Equation (2.1) distance d is squared because free space is considered. In practice, this has to be replaced by d raised to power 'n' where 'n' (some literature also uses 'ζ' (zeta)) as the path loss exponent. It is essential to ensure that $P_t$and P_r are in the same units, while G_t and are gained (unit less). Miscellaneous losses denoted as L $(L \geq 1)$ typically stem from attenuations present in transmission lines, losses in filters and antenna losses within the communication system. Note that $L = 1$ signifies no hardware losses in the system.

The Free Space Path Loss Equation (2.1) illustrates that the received power decreases by the square of the transmitter-receiver (T-R) distance. This indicates that the received power diminishes with distance at a rate of 20 $dB/decade$. The antenna gain is linked to its effective aperture(A_e), by

$$G = \frac{4\pi A_e}{\lambda^2} \tag{2.2}$$

A_e is connected to the physical antenna dimensions, and λ is linked to the frequency by

$$\lambda = \frac{c}{f} = \frac{2\pi c}{\omega c} \tag{2.3}$$

When discussing carrier frequency, f represents it in Hertz, w_c in rad/sec, and c as the speed of light in m/sec. An isotropic antenna emits power with a uniform gain in all directions. It serves as a standard for comparing antenna gains in wireless systems. When compared to an isotropic radiator, the Effective Isotropic Radiated Power (EIRP) is the highest directional radiated power that can be produced by a transmitter and it is defined as:

$$EIRP = P_t . G_t \tag{2.4}$$

In practical applications, Effective Radiated Power (ERP) is utilized in place of EIRP to indicate the maximum radiated power of a half-wave dipole antenna rather than an isotropic antenna. In general, these figures are denoted in dB because the products (in traditional form) form become summations in dB scale. Antenna gains can be presented in dBi, the gain concerning the isotropic source. Non-isotropic antennas, such as directional antennas, can have better gains, but the gain relative to the isotropic antenna can be better represented as dBi. For example, a cell-based antenna may have a gain of 10 dBi. Path loss, which indicates attenuation value measured in dB, is defined as the discrepancy between the P_t and P_r, and also considers the influence of antenna gains. The path loss in the free space model, accounting for antenna gains G_t andG_r, is provided by:

$$PL(dB) = 10\log[\frac{Pt}{Pr}] - 10\log[GtGr \quad \frac{\lambda^2}{(4\pi)^2 d^2}] \tag{2.5}$$

Suppose, antennas are considered to have gain equal to 1 at T-R both sides, then antenna gains are eliminated from above formula, and PL is determined by

$$PL(dB) = 10log[\frac{Pt}{Pr}] = -10log[\frac{\lambda^2}{(4\pi)^2 d^2}] \quad (2.6)$$

The Friis free space model can only accurately forecast P_r for all possible values of 'd' that fall inside the transmitter's far-field region. The region beyond the far-field distance df that is connected to the carrier wavelength and the largest dimension of the Tx antenna aperture is known as the far-field, or Fraunhofer region, of a transmitting antenna. One can find the Fraunhofer distance using,

$$\text{df} = \frac{2\text{D}^2}{\lambda} \quad (2.7(a))$$

Where, D is the antenna's greatest physical dimension. Furthermore, for df to be in the far-field zone, it needs to fulfill the following equation,

$$df \gg D \quad (2.7(b))$$

and

$$\text{df} \gg \lambda \quad (2.7(c))$$

Further it can be noted that Equation (2.1) is not applicable when, $d = 0$. Due to this limitation, propagation models adopt 'do' as a reference point for received power. The received power, (d) , at distances greater than do, can be correlated to $P_r(do)$ which can be estimated from Equation (2.1) or computed by measuring the average of received power levels at multiple points close to the transmitter at a radial distance of do in the radio environment. The reference distance must be carefully selected so that, $do > df$. Additionally, 'do' should be smaller in comparison to distance used in mobile communication systems. Therefore, based on Equation (2.1), the P_r in free space at a distance 'd' beyond do is expressed as:

$$Pr(d) = Pr(do)(\frac{do}{d})^2 \qquad d \geq do \geq df \quad (2.8)$$

In practical mobile environments, very often we observe significant changes in P_r by many orders of magnitude across a typical coverage area spanning a few square kilometers. Received power levels, measurements are often expressed in dBm or dBW units. Equation (2.8) can be converted into units of dBm or dBW as,

$$Pr\ \Pr(d) = 10\ log\ \log\left[\frac{Pr\ Pr\ (do)}{0.001W}\right] \quad (2.9)$$

$$+\ 20\ \log\left(\frac{do}{d}\right)[in\ dBm] \qquad d \geq do \geq df$$

Where, P_r (do) is measured in watts. Note that received power is normalized with respect to 0.001 W. In practical systems utilizing low-gain antennas in the 1-2 GHz range, the reference distance do is commonly set at 1 meter in indoor settings and either 10 meters or 1 kilometer in outdoor settings. This will ensure path loss calculations in dB units.

Example 2.1: Consider a base station transmitting with power of 10 watts, frequency 900 MHz and transmitting antenna gain equal to 2. Consider a down link case whereby BS is radiating power and the user is trying to receive signals at MS. Receiver gain is 1 and MS is currently located 5 km away from BS. Calculate Received power in dBW and dBm.

Solution: Given: for a BS , P_t= 10 W , $G_t = 2\ and\ G_r = 1$, d=5000 m

$$\lambda = \frac{c}{f} = \frac{(3*10^{8})}{(9*10^{8})} = 0.33 \text{ m}$$

$$P_r(d) = 10log\left\{\left(\frac{P_tG_tG_r\lambda^2}{4\pi}\right)^2 d^2\right\} = 10log\left\{\frac{10\times2\times1\times(0.33)^2}{(4\pi)^2} \times (5000)^2\right\}$$

$$P_r(d = 5000m) = -92.6\ dBW = -62.6\ dBm$$

Example 2.2: Consider another case with a base station transmitting with power of 500 m watt. Rest of the parameters are kept the same. MS has moved out to a distance of 10 km away from BS. Calculate Received power in dBW and dBm. It is given that GSM receiver system (MS) sensitivity is -100 dBm. Question is will the system work properly and receive signals?

Solution: Given data is same as earlier examples with changes $Pt = 500$ mw and d=10 km.

$$P_r(d) = -10 \; log \; log \left[\frac{P_t \; G_t \; G_r \lambda^2}{((4\pi)^2 d^2 \;)}\right]$$

$$= 10log[(0.5 \times 2 \times 1 \times \frac{0.33^2}{4\pi^2} \times (10000)^2]$$

$$P_r(d = 10000m) = -111.6 \; dBW \; = -81.6 \; dBm$$

It is clear that received power is higher than minimum sensitivity of MS and hence, the system will properly work and will receive signals.

2.1.2. Understanding Basic Radio Propagation Mechanisms

In order to design any cellular mobile system, power control and power monitoring are among important parameters. However, when BS is receiving an attenuated signal it becomes important to understand how exactly the signal transmitted reaches the BS antenna. Attenuated signal may reach the receiver either via LOS or via NLOS having few reflections or through diffraction or through scattering. One of the aspects is signal coming via reflections.

Reflections can be categorized like from a metallic source or it could be a non-metallic material like wood, concrete building or any other thing. Another possibility is a signal from NLOS through scattering. Another possibility is diffraction across the edges. Again, all these parameters will change as frequency bands change. All these mechanisms have their own problems. Considering propagation issues designers have to identify their sources (a) reflections or (b) diffraction or (c) scattering or combination of all three.

When an electromagnetic wave strikes any particular object larger than the incident wave's wavelength, such as the plain Earth surface, walls of large buildings or similar surfaces, reflection takes place. At least an order of magnitude difference is usually present for large objects, and reflections can happen at wavelengths of 0.33 m for 900 MHz and 0.175 m for 1800 MHz.

On the other hand, when there is any sharp irregularity, such as an edge, in the path between the T-R, diffraction occurs. This clarifies how radio signals can pass through both rural and urban areas without requiring a direct line of sight. When there are smaller or comparable objects to the wavelengths in the medium, such as dust particles, raindrops, snowflakes, and rough surfaces, scattering takes place.

2.2. Noise and Interference

Unwanted electrical signals that obstruct intended signals are called noise. It originates from both artificial and natural sources, including communications systems, commutator sparking, and car ignition circuits. Artificial sources result from harmonics of natural frequency, while natural sources include circuit noise, atmospheric disturbances, and extraterrestrial radiation.

2.2.1. Thermal Noise

One essential characteristic of matter above $0°K$ is thermal noise. Electrons are bonded to the atom in an insulator and free to travel in a conductor. Free electrons travel randomly in a conductor above $0°K$, averaging zero velocities over time. Spontaneous fluctuations, which are analogous to intermittent currents, happen over brief periods of time. For instance, thermal noise causes the voltage between a metallic resistor's two terminals to be nonzero. This mean-square voltage ($\tilde{v}$) is given as,

$$\tilde{v} = \left[\frac{2\pi^2 k^2 T^2}{3h}\right] \times R \tag{2.10}$$

Where,

R - represents resistance (ohms)

Boltzmann's constant, $k = 1.37 \times 10^{-23}$W-s/°K

Planck's constant, $h = 6.62 \times 10^{-34}$W-s^2

T is the absolute temperature in $°K$.

Thus, the resistor can be modeled as the equivalent of the Thévenin circuit depicted in Figure 2.1, where the voltage source has a mean-square voltage of Equation. (2.10) and a zero mean for the resistor, which is ideal (noiseless).

The voltage can be accurately modeled as a Gaussian distribution with zero mean as it is the sum of the motions of many electrons. The resistance R determines the mean-square (m.s.) voltage across the resistor terminals, which is finite. When the load resistance is

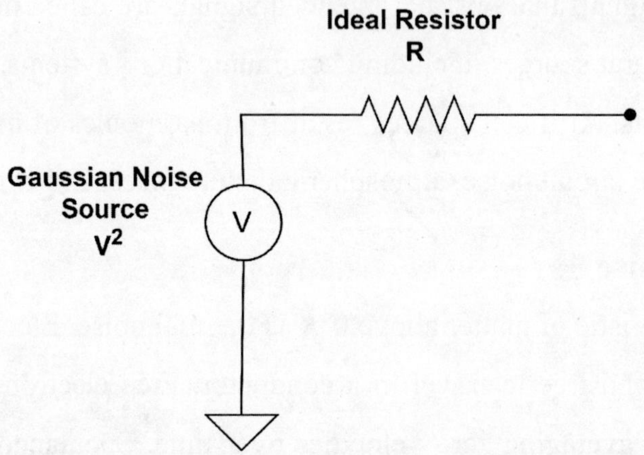

Figure 2.1 Thévenin model of a resistor.

equal to R, the circuit in Figure 2.1 delivers its maximum power. An equation for the maximum achievable noise power can be obtained by scaling by a factor of $\frac{1}{4}$ and dividing both sides of Equation. (2.10) by R.

A relatively small fraction of the entire electromagnetic spectrum is used by any communications link. The distribution of thermal noise power across frequency, $S_n(f)$or the noise spectral density, is of relevance to communications engineers. This is also expressed in terms of quantum physics, with the key finding being that the available thermal noise spectral density (watts/Hz) is about constant at frequencies up to 10^{12} Hz.

$$S_n(f) = \frac{kT}{2} \tag{2.11}$$

$$\equiv \frac{No}{2} \qquad -\infty < f < \infty$$

The noise power applied to a matched load is described by Equation (2.11); it is independent of resistance. The highest limit of 10^{12} Hz is located well above the point at which traditional electrical components stop responding, in the near-infrared region of the electromagnetic spectrum. The noise power falls down exponentially with frequency above these frequencies.

Fig. 2.2 shows the two-sided spectrum model for thermal noise, which is employed based on the Equation (2.11). This model demonstrates that thermal noise has density $\frac{No}{2}$ and a flat spectrum across all frequencies, both positive and negative.

Given that all of the frequencies are present at the same volume, this kind of noise is known as white noise. At all frequencies, the amplitude distribution is Gaussian, and samples of the corresponding time-domain noise process $n(t)$ do not correlate with each other, which means, the autocorrelation function for a genuine white-noise process is provided by:

$$R_n(\tau) = E[n(t)n(t+\tau)] \qquad (2.12)$$

$$= \frac{\left[\frac{No}{2}\right]}{\delta(\tau)}$$

Conventionally, the noise spectral density is described as a two-sided number $\frac{No}{2}$ that is valid across the range $-\infty < f < \infty$.

The power of noise in bandwidth B is

$$\mathrm{P} = \left(\frac{\mathrm{No}}{2}\right)(2\mathrm{B}) \qquad (2.13)$$

$$= \mathrm{N_O B}$$

Where, two- sided representation is represented by the first equation and the single-sided representation by the second. When a filter exhibits a frequency response $H(f)$, its noise-equivalent bandwidth, or noise bandwidth, is defined as follows:

$$B_{eq} = \frac{\int_{-\infty}^{+\infty} |H(f)|^2 \left(\frac{N_o}{2}\right)^2 df}{N_o/2} \tag{2.14}$$

The noise power bandwidth is the the bandwidth over which the same amount of noise power would travel through an ideal rectangular filter. The noise bandwidth of many filters is roughly equivalent to their 3-dB bandwidth. Similar methods can be used to define a signal's noise-equivalent bandwidth.

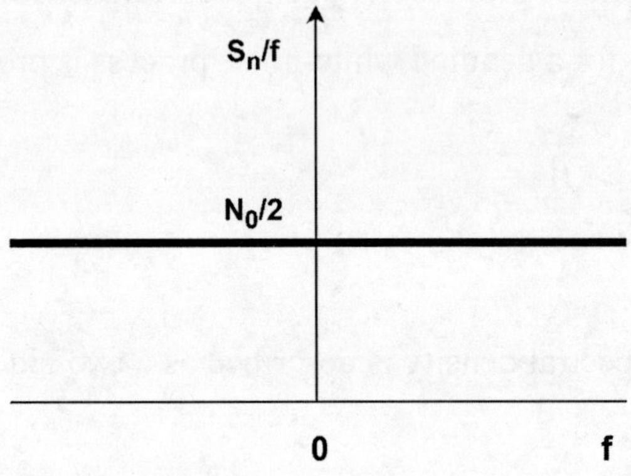

Figure 2.2 White noise spectral model.

2.2.2. Equivalent Noise Temperature and Noise Figure

There are other forms of noise that behave similarly to thermal noise, implying that their noise is white across the relevant frequency range and has a Gaussian distribution. Because of this, these sources are frequently associated with the idea of an analogous noise temperature. Instead of measuring a device's actual temperature, the equivalent noise temperature is a measurement of the noise the gadget makes. While transmission amplifiers do add noise and that is represented as noise figure, which is given by:

$$F = \frac{S_{no}(f)}{G(f)S_{ni}(f)} \tag{2.15}$$

Where, $S_{no}(f)$ is the output noise power spectrum, $G(f)$ is available power gain of a device (it is depending frequency) and $S_{ni}(f)$ is the input noise power spectrum. In an ideal noiseless amplifier, the noise figure is unity. That is where, T_0 is the standard temperature of 290 degrees Kelvin. The dependency between the equivalent noise temperature (T_e) and the noise figure (F) is given by expression:

$$T_e = (F - 1)T_0 \tag{2.16}$$

The noise power due to the receiver in a bandwidth B is given by:

$$\mathrm{N = F(kT_0)B} \tag{2.17}$$

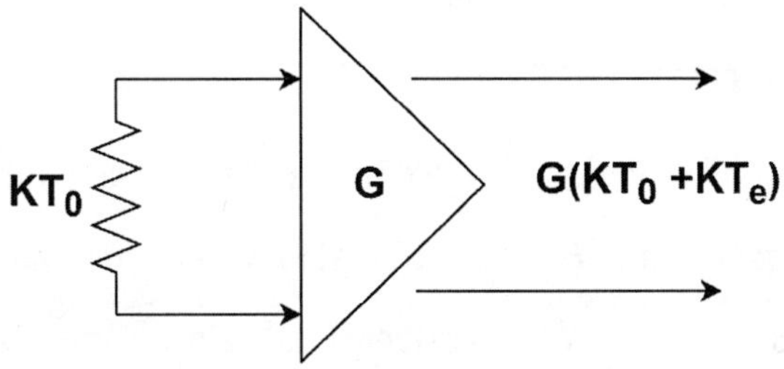

Figure 2.3 Noise Model of amplifier

2.2.3. Noise in Cascaded Systems

Transmission lines connect the many parts that make up a communications receiver, which also includes an antenna, amplifiers, filters, and mixers. These components contribute noise in a weighted manner to the overall system temperature.

All the devices indicated above are modeled using the two-port model presented in Fig. 2.3, where each device has a gain (or loss) G and a noise factor F. In this model, an internal noise source is added to the device's input. The internal noise source's spectral density is $(F - 1)kTo$. Following that, an ideal (noiseless) amplifier with gain G processes the combined signal. When $S(f)$ is used as the device input, the circuit's output may be found by

$$Y(f) = GS(f) + G(F - 1)kTo \tag{2.18}$$

This output has both a signal component and a noise component.

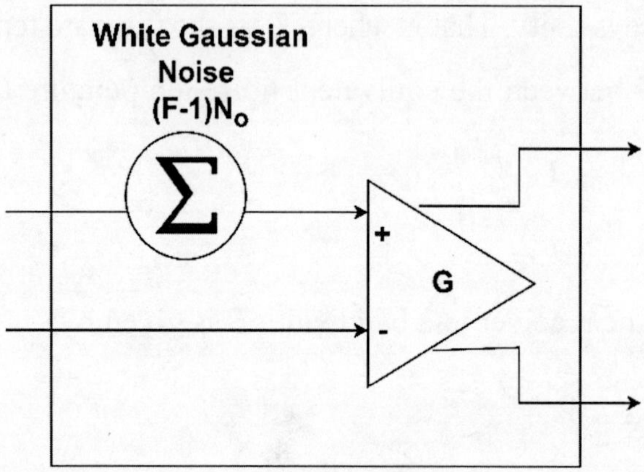

Figure 2.4 Two-port model for a communications system component.

As shown in Figure 2.4, now consider a system with two such devices, representing gains $G_1$and G_2 and noise factors F_1 and F_2respectively. We assume that the input is linked to a metallic resistor with noise density N_0 in order to calculate the noise figure of the cascaded system.

Based on these presumptions F_1 $G_1 . N_0$ is the first stage's output the same as the second stage's input. G_2 is the second stage's output. According to the definition of noise figure provided in Equation 2.18, the overall noise figure is obtained by comparing the cascaded devices' input and output. This is defined as follows:

$$\mathrm{YF} = \frac{[\mathrm{G_2\{(F_2-1)N_0+F_1G_1N_0\}}]}{\mathrm{G_1G_2.N_0}} \quad = \mathrm{F_1} + \left[\frac{\mathrm{F_2-1}}{\mathrm{G_1}}\right] \tag{2.19}$$

In other words, the combined noise figure is equal to the sum of first device's noise figure plus the second device's noise figure, attenuated by the first device's gain. Typically, a multistage system's noise figure is provided by

$$F = F_1 + \left[\frac{F_2-1}{G_1}\right] + \left[\frac{F_3-1}{G_1 G_2}\right] + \ldots. \tag{2.20}$$

Consequently, the system's equivalent noise temperature is

$$T_{sys} = T_1 + \left[\frac{T_2}{G_1}\right] + \left[\frac{T_3}{G_1 G_2}\right] + \ldots. \tag{2.21}$$

Hence, for each k in the receiver amplification chain, T_k and G_k represent the noise temperature and gain of the kth stage, respectively. When we incorporate the antenna's equivalent noise temperature into Equation (2.21), the formula changes to

$$Tsys = T_\alpha + T_1 + \left[\frac{T_2}{G_1}\right] + \left[\frac{T_3}{G_1 G_2}\right] + \ldots \tag{2.22}$$

Where, T_α is the equivalent noise temperature of the antenna.

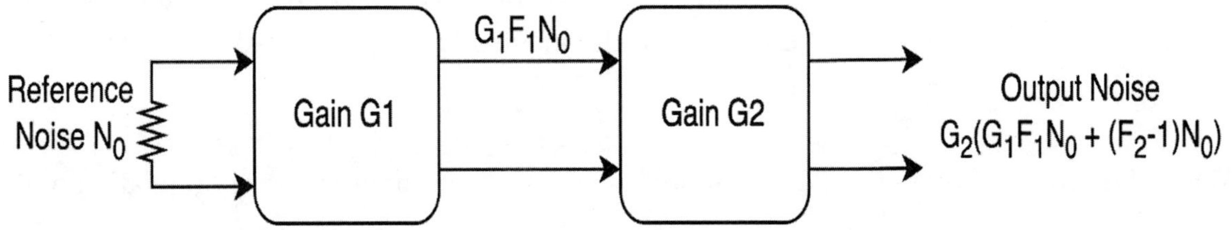

Figure 2.5 Calculation of noise for cascaded system

2.2.4. Man-Made Noise

The internal and external noise sources have an impact on digital communications systems. Good system and antenna design can reduce internal and external noise, whereas artificial noise levels can set a receiver's sensitivity limit. Large electrical discharge devices, electromagnetic pulses from electrical machinery, spark ignition systems, switching transients, and discharge lighting are three examples of artificial noise sources. A non-stationary in nature, impulse noise is influenced by various elements like operating frequency, time of day, and surrounding environment. Measurements in the frequency band of interest, in various settings, and at various times of the day, week, and year are necessary for accurate rate modeling. Interference can also be caused by out-of-band emissions from other services that resemble communications, like television signals and powerful radars.

2.3. Three Basic Propagation Mechanisms

In practice, mobile networks are being influenced by the three basic mechanisms (a) reflection, (b) diffraction, and (c) scattering. Design of propagation models mainly use these three propagation concepts to forecast received power or path loss. An electromagnetic wave will reflect off of walls, buildings, and the surface of the earth when it comes into contact with an object that is substantially larger than the wave's wavelength. When a surface with sharp edges or other irregularities obstructs the electromagnetic route between the transmitter and receiver, diffraction takes place. Even in situations when there isn't a direct LoS between T-R, the secondary waves caused by the blocking surface are present across the area and even behind it, causing the waves to bend around it. Similar to reflection, diffraction at high frequencies is dependent on the geometry of the object and the incident wave's amplitude and phase at the point of diffraction.

When the wave's medium is made up of objects whose dimensions are small in relation to the wavelength and such obstacles density is high, scattering takes place. Rough surfaces, tiny particles, and other imperfections in the channel can cause scattered waves. In actuality, a mobile communications system experiences scattering due to trees, traffic signs, and lampposts.

2.3.1. Reflection

Let us understand with more details the working of reflection mechanisms for propagation of radio signals. Reflection, as defined earlier, occurs when a propagating wave moves from one medium to another medium that has different electrical properties. Different kinds of reflection will occur in accordance to different electrical properties in the dielectric or on a conductor.

When radio waves interact with various electrical characteristics of a material, they are partially reflected and transmitted. All incident energy will be reflected back into the first (original) medium without any energy loss if the second medium is a perfect conductor. The Fresnel reflection coefficient (Γ) allows the electric field intensity of reflected and transmitted waves to be connected to the incident wave in the medium of origin. It is important to note that good metal conductors like windows, building tops, surfaces which have metallic surfaces will form good reflectors.

2.3.1.1. Reflection from dielectric

The reflection of an electromagnetic wave at an angle of Θ on two dielectric mediums is shown in Figure 2.6. It can be noted that the energy is propagated into the second medium and/or reflected back to the first. The wave is broken up into two orthogonal components E_i andH_i. Out of the same, E_r and H_r components are reflected E_T is transmitted. The type of reflection depends on the E-field's direction of polarization. There are two different examples shown in Fig.3.4, where the E-field polarization is either perpendicular to the plane of incidence (perpendicular to the reflecting surface) or parallel to the plane of incidence (vertical). One can examine this behavior for any direction of polarization.

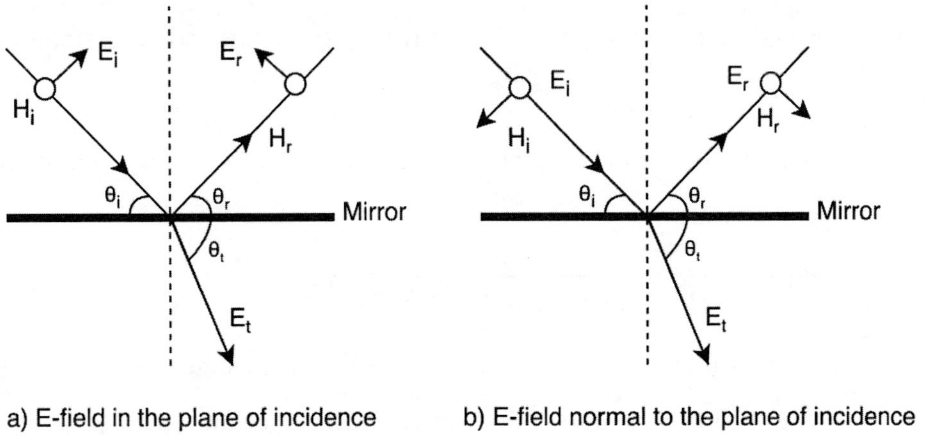

Figure 2.6 Geometry for calculating the reflection coefficients between two dielectrics.

2.3.1.2. Ground reflection (2-ray) model

Transmission across mobile radio channels does not always depend on a direct path between a mobile device and the base station. As a result, it is frequently incorrect to rely only on the free space propagation model. Based on geometry, the 2-ray ground reflection model (see Figure 2.7) provides useful insights to the propagation. It takes into consideration both the transmitter and receiver's direct path as well as a path reflected by the earth. This model is known to be reasonably accurate in estimating signal strength for mobile networks using tall towers (over 50 meters in height) and for LOS in urban environments over long distances (several km). Hence, the ground reflection model is a method used to study the transmission of signals between a BS and a MS. The BS, typically 50 meters or higher, is erected on the peak of a tower or small structure on terrace of a building. The model assumes that the MS is located at a certain distance away from the BS. If this criterion does not match, the model may not work properly.

When a communication is established between BS and MS; the energy received (E_{TOT}) is the sum of the LOS path (E_{LOS}) and the reflected path (E_r), but this is not a scalar sum but a vector sum due to the path difference. The height of the base station (h_t) and the receiver (h_r) also play a role in the received power. The distance (d) between the transmitter and receiver can be taken from the ground or from the ground level if the BS is situated on a small hill or a tall building.

The path difference (Δ) between direct LOS and the reflected paths from ground can be computed from geometry as,

$$\Delta = \sqrt{(h_t + h_r)^2 + d^2} - \sqrt{(h_t - h_r)^2 + d^2} = 2\frac{h_t h_r}{d} \tag{2.23}$$

In above equation, h_t is fixed whereby h_r may vary. When the path difference is known, the following relations can be used to compute the phase difference, θ,

$$\theta = \frac{2\pi\Delta}{\lambda} = \frac{\Delta\omega}{c} \tag{2.24}$$

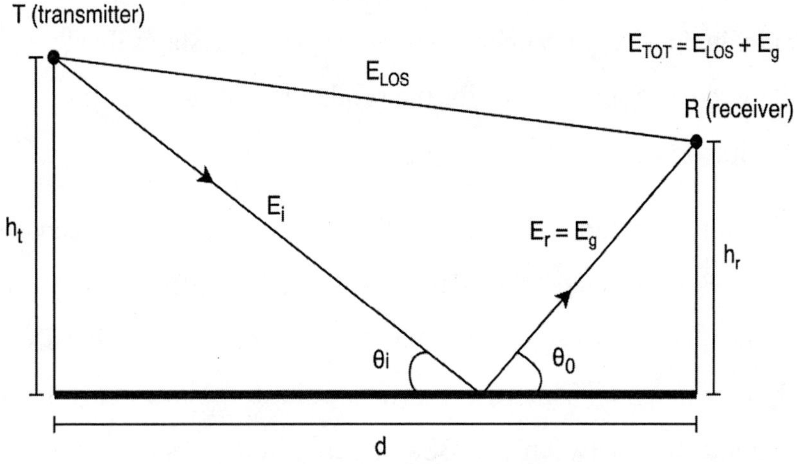

Figure 2. 7 Two-ray ground reflection model

Phase difference θ depends on the frequency and the wavelength. The time delay (τd), time difference between arrival of the two paths is given as,

$$\tau d = \frac{\Delta}{c} = \frac{\theta}{2\pi f} \tag{2.25}$$

The power received(P_r) (distance d from Tx) is computed as:

$$P_r = P_t\, G_t\, G_r (\frac{{h_t}^2 {h_r}^2}{d^4}) \tag{2.26}$$

This is a very simplistic and realistic model and generally performs well at the GSM frequencies. However, it should be noted that as the distance between T-R increases, the LOS condition gets worse. It becomes more and more difficult to find LOS as MS continues its journey away from BS. Designers have to keep all these factors in mind before applying models in practice.

2.3.2. Diffraction

Diffraction is a phenomenon where a sharp irregular surface obstructs the radio path between the T-R, causing a loss of signal. Various components like edges, corners, bends, etc. cause diffraction. Concept of diffraction explains reception of mobile signals in a room or office environment that are not in direct line of sight from BS. This is crucial as otherwise, radio waves cannot travel in dense environments. Huygens principle states that "all points on a wave front can be considered as point sources for the production of secondary wavelets" and can be used to explain diffraction.

Radio signals can travel without a direct LOS, a phenomenon that can be explained by diffraction models. For instance, consider 'a knife edge diffraction geometry' with the presumption that there is a sharp edge at the top. Consider a tower with a Tx antenna and one more tower with a Rx antenna. This case can be a point to point microwave communication link or as a communication between two BSs or between the BS and MSC. The first tower's signals must radiate towards the receiver antenna, which takes in the energy. The wave propagates in all directions as it is sent from the transmitter, and many secondary wavelets are produced at any point along the wave front. Despite the lack of a LOS, the receiver receives the energy from the diffraction.

In Figure 2.8 the illustration of knife-edge diffraction geometry. The receiver R is located in the shadow region. It should be noted that from the transmitter side d_1 is the distance to the knife edge obstruction and from the receiver distance is d_2. Also important will be the height (h) of the obstruction. From these parameters energy at the receiver can be computed.

The Frensel-Kirchoff diffraction parameter is defined as follows:

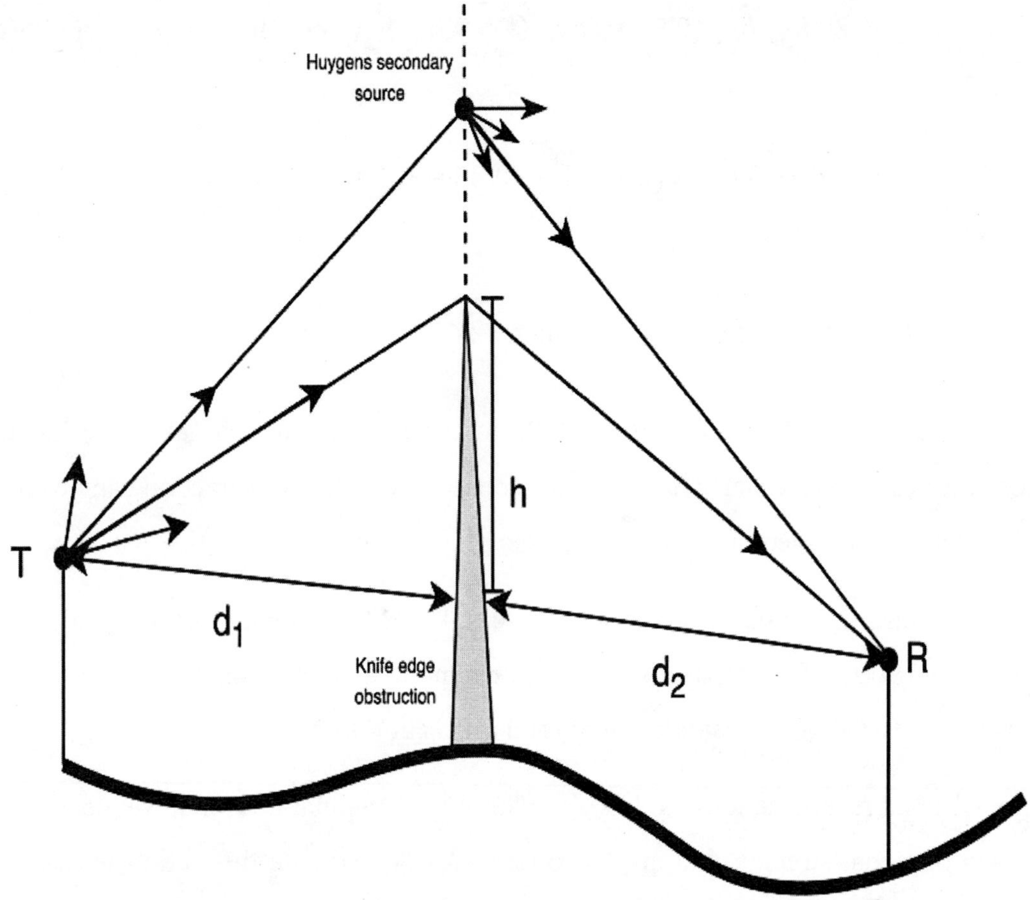

Figure 2. 8 Illustration of knife-edge diffraction geometry

$$v = \frac{h\sqrt{[2(d_1+d_2)]}}{\lambda d_1 d_2} \tag{2.27}$$

Once diffraction parameter (v) is obtained for a given geometry, a knife edge's diffraction gain (G_d) is calculated using,

$$G_d(dB) = 0 \quad \text{when } v \le -1 \tag{2.28(a)}$$

$$G_d\,(dB) = 20\,log\,log\,(0.5 - 0.62v) \quad \text{when } -1 \le v \le 0 \tag{2.28(b)}$$

$$G_d\,(dB) = 20\,log\,log\,(0.5\,exp\,exp\,(-0.95v)\,)\ \text{ when }\ 0 \le v \le 1 \qquad (2.28(c))$$

$$G_d\,(dB) = 20\log(0.4 - \sqrt{(0.1184 - (0.38 - 0.1v)^2})\ \text{ when } 1 \le v \le 2.4 \qquad (2.28(d))$$

$$G_d\,(dB) = 20\,log\,log\,\left(\frac{0.225}{v}\right)\ \text{ when }\ 2.4 \le v \qquad (2.28\,(e))$$

Above equations 2.28 (a-e), suggest formula to compute diffraction gain (G_d) for different values of diffraction parameter (v). Energy to be received after a diffraction can be computed. However, it is to be noted that wavelength also plays its role.

This means that the same building will have 2.4 GHZ frequency band signal diffraction different from 900 MHz. Propagation pathways in mountainous terrain frequently have several obstacles, necessitating the calculation of total diffraction loss.

Single knife-edge diffraction models oversimplify computations and frequently produce estimates of signal strength that are too optimistic. To estimate diffraction losses owing to various obstructions, numerous simpler models have to be considered.

2.3.3. Scattering

When there are tiny particles, rough surfaces, raindrops, plants, or other objects smaller than or equivalent to the wavelength present in the medium, scattering takes place. Foliage gains importance as frequency increases (or wavelengths decrease). There aren't many significant obstacles to GSM frequency propagation via trees at 900 MHz and 800 MHz. Green patches, on the other hand, cause diffraction, scattering, and absorption at frequencies greater than 10 GHz. Similar to diffraction, scattering causes transmitter energy to be emitted in all directions, including at edges, foliage and street signs.

2.3.4. Some Scenarios of Radio Mechanisms

In real-life scenarios, the system planner may place a BS on top of a building to cover the maximum area. However, such optimal BS location may not be available for purchase, leading to sub-optimization. This is an optimization problem and resource allocation problem. For BS location management, optimization problems can be solved as what is the minimum BS terminals required to satisfactorily serve a given geographic area? A mobile receiver with a small antenna can move into the different environment conditions. The BS antenna height and the MS antenna height are important factors to consider.

Radio propagation mechanisms in real-life urban scenarios involve tall buildings, towers, ground, and a mobile station. The direct signal has a direct line of sight, but it also has a reflected signal. This can cause destructive interference and fading, which will be discussed in the next unit. Reflected signals can also be diffracted from an edge. A net signal received due to reflections and diffractions may be very different from what was actually sent. This is due to the fact that all paths have different time delays, different phases resulting in different signal strengths. T-R distance is crucial in understanding propagation mechanisms.

Let us consider a scenario from a bird's eye perspective, an aircraft / drone is observing a city with buildings under the street. Tx antenna is placed in a central space, with a small roundabout where a police person can stand. The transmitting antenna communicates with another mobile station, falling under the policy of wireless communication. The receiver, free to move, moves in pedestrian traffic, losing line of sight. However, the receiver still receives energy through reflections and diffraction. In this scenario, the receiver is not a reflection, but a knife edge diffraction. This diffraction occurs at a corner of the building and the person still receives energy.

Scattering effects from this point depend on the wavelength being used. In this simple scenario, reflection, diffraction at two places, and an edge playing an important role. All three factors are playing an important role in the propagation mechanism. Therefore, the propagation mechanism in a city with buildings under the street is influenced by reflection, diffraction, and scattering.

Hence, the instantaneous received signal strength is affected by three mechanisms when a mobile device moves across a service area. Diffraction and scattering won't be the main modes of propagation if a mobile device has an unobstructed line of sight. Diffraction and scattering might be more prevalent, though, if the mobile device is at street level and not in a line of sight. This is significant since there are models that calculate the real received power and the feasibility of doing the work well.

2.4. Radio Propagation Model

2.4.1. Introduction

A propagation model is essential for calculating the coverage area of a transmitter and the cell size. A good model properly forecasts cell size based on signal power, which determines transmit power requirements and battery lifetime. System designers can increase channel quality by taking into account a proper propagation model and determining appropriate modulation and coding schemes. A pessimistic propagation model may not forecast adequate signal intensity, resulting in weaker modulation methods such as BPSK or QPSK.

However, a more accurate channel and propagation model may result in increased signal strength and modulation techniques, hence enhancing data rate.

2.4.2. Free Space Propagation Model

We have already seen details of the free space propagation model in earlier sections. However, for continuity we shall revisit important equations again.

Friis free space equation is defined as:

$$(d) = \frac{P_t G_t G_r \lambda^2}{(4\pi)^2 d^2 L} \tag{2.29}$$

The Friis free space equation states that power decreases with distance between transmitter and receiver, and received power decays at a rate of 20 dB per decade.

Power received at a distance $d \geq d0$ id defined as

$$P(d) = (d_0)\left(\frac{d_0}{d}\right)^2 \tag{2.30}$$

Above equation (2.30) is valid for a condition defined as

$$d > d_0 > d_f \tag{2.31}$$

Where, d_0 is a reference distance and d_f is the Fraunhofer distance.

Power level in dBm is given by:

$$P(d) = -10\,log\left(\frac{(d_0)}{0.001}\right) + 20\,log\left(\frac{d_0}{d}\right) \quad \text{(dBm)} \qquad d > d_0 > d_f \tag{2.32}$$

Received power in practical scenarios can be calculated by:

$$P_r = P_t G_t G_r \left(\frac{h_t^2 h_r^2}{d^4}\right) \tag{2.33}$$

The ideal situation for this calculation is without interference, reflection, or scattering, which is valid for satellite communication. Different models, derived from measurements, curve fitting, or theoretical analytical work, can be used to account for these effects. Despite the limitations of this model, it can be used to predict receiver strength for mobile networks.

2.4.3. Introduction to Radio Propagation Models

It is necessary that designed signal models should be able to mimic the signal strength correctly that reaches the receiver at any given location after having effects of different propagation mechanisms like reflections, diffractions, and scattering. Factors like relative locations of T-R, actual number of reflectors, all affect the amount of reflection, diffraction, and scattering. Propagation models on small and big scales are distinct from one another. Radio propagation models can be created analytically by mathematically modeling propagation mechanisms and deriving path loss equations, or empirically by gathering measurement data and fitting curves. These realistic models can be applied to a variety of situations, including rural, vegetated, dense, and uneven terrain.

2.4.4. Small-Scale Propagation Model

Mobile devices moving over short distances cause rapid fluctuations in instantaneous received signals, which can be studied using small scale propagation models. It should be noted that here small distances are defined on scale nearer to the wavelength (lambda) or frequency (f). These models undergo quick fluctuations due to the algebraic sum of several contributions from various directions, such as scattering or diffractions or reflections. Received signal at a given time is due to the arrival of multiple replicas of the same transmitted signal that are coming with different phases and different time. As phases of these signals are random, signals can be considered to have homogeneous distributions with random phases between 0 and 2pi. When combined, these elements exhibit noise-like behavior and can be represented with random variables e.g. Rayleigh fading. The received signal power can fade very much during small-scale fading, producing an extremely low received power. It should be noted that these 30-40 dB change in signal strength happened with a MS movement of only a fraction of wavelength. Hence, a fading can lead to extremely low received power for few instances only and hence should not be the sole foundation of a handoff strategy. Nevertheless, the receiver exits the fade when it shifts a portion of the wavelength or the majority of the wavelength. Thus, receiver diversity can be achieved if two transmit antennas are positioned so that one may be in the fade and the other may not be.

Small transmitted receiver (T-R) separation distance variations, which are commonly observed in metropolitan environments with densely occupied buildings, scatterers, and powerful reflectors, determine the properties of small scale propagation models. The primary method of signal transmission is scattering, in which several copies of the sent signal arrive at the receiver by receiving paths and contribute vectorially to the receiver at varying times, resulting in fading. Considering the presence or absence of LOS, the signal attenuation constant's distribution can be either Rayleigh or Rician.

2.4.5. Large-Scale Propagation Model

A model that is able to forecast average power levels for any T-R separation over a larger distance is known as a large-scale propagation model. It covers the entire cell at long transmitter-receiver separation distances. With a limited number of reflections, reflection serves as the primary propagation mechanism. Shadowing is used to describe signal attenuation caused by power loss along the path traveled, and log normal distribution can be used to demonstrate the power loss distribution in dBs. Large-scale propagation requires a log normal shadowing model. Unlike the rapid fluctuations around a slowly varying mean for small scale propagation models, there are minor fluctuations around a slowly varying mean. This is utilized for cell site planning and is helpful in determining a transmitter's radio coverage area.

By conducting measurement campaigns, averaging the power received at a distance of 4 to 5 wavelength movements of MS with respect to its BS, and plotting the data to produce the requisite propagation curve, a suitable large-scale propagation model can be developed. Moreover, to depict the power loss in dBs, the log normal type of distribution for power loss is used.

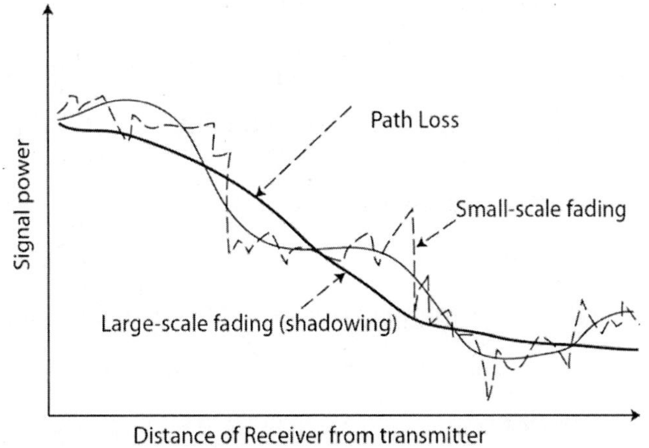

Figure 2.9 Large and small-scale variations in signal strength over time

2.4.6. Log-Distance Path Loss Model

This type of model is a tool used to calculate the average large scale path loss for an arbitrarily separated T-R separation. This model uses loss exponent 'n' to characterize the propagation environment, with values ranging from 2 to 6 depending on the obstructions present. The relationship between path loss and distance is based on measurements and empirical modeling. This can be expressed as Equations 2.34 (a-b):

$$\underline{PL} \propto \left(\frac{d}{d_o}\right)^n \ldots\ldots\ldots d \geqslant d_o \qquad (2.34)(a))$$

$$\underline{PL}\,(dB) = \underline{PL}\,(d_o) + 10n * log\left(\frac{d}{d_o}\right) \qquad (2.34(b))$$

It should be noted that path loss exponents for large scales typically vary across types of environments, with free space having a value of 2. However, in urban areas, due to more obstructions, with values ranging from 2.5 to 3.5. In the city area a value close to 3.5, while in the suburbs area a value closer to 2.5. Shadowed urban areas can have values as high as 3.5 (or more), as there is no line of sight in buildings. Building line of sight can range from 1.6 to 1.8, as there is a strong guiding effect in buildings and corridors, which helps to reduce energy loss. Even narrow streets can give a value close to 2, making it a better choice for certain applications. Therefore, the log distance path loss model provides valuable insights into the path loss exponents for various propagation environments.

By understanding the factors that contribute to the path loss exponents, researchers can develop more effective strategies for improving their signal quality and performance.

2.4.7. Log Normal Shadowing Model

Log normal shadowing is a concept that has been proposed to reflect the shadowing effects due to presence of cluttering on the propagation path. The standard path loss model equation generally does not consider the surrounding environment. This may go worse for measurements for long distances with two different Tx-Rx locations having the same distance. In such cases, predicted average power values will be quite different than actual.

Measurements show that for received value at a location'd', the path loss PL (d) in dBm at a location is randomly distributed log normally. The log normal shadowing equation typically measures and predicts the average path loss as a function of "d" - the T-R separation and the path loss exponent "n". However, the environment may be different, and there must be randomness. To address this issue, a component of randomness must be added to the Equation 2.34 (a-b) to get new Equations as 2.35 (a-b).

$$\underline{P}L\ (dB) = \underline{P}L\ (d) + X_\sigma \tag{2.35(a)}$$

$$\underline{P}L\ (dB) = \underline{P}L\ (d_o) + 10n\ log\ log\ \left(\frac{d}{d_o}\right) + X_\sigma \tag{2.35(b)}$$

It has been proven that path loss measurement at various distances can be shown as a distribution. A correction factor, X_σ (0) that is a 'zero mean' Gaussian with a standard deviation SD_σ in dB. This randomness is added into the path loss equation, and the received power is calculated from measured data. Value of sigma (σ) can be derived from either measurements or analytically. In applications, both exponent (n) and SD_σ are calculated from measured data. Once completed, the received power in dBm equals Tx power (in dBm) minus PL. In a suburban setting, a model designer records received power values while moving and takes multiple measurements, averaging them to determine the average received power. Initially, the reference power at a specific distance, like 100 meters from BS, is established. The received power is then calculated relative to 1 mW. Moving outward from BS, measurements are taken at different intervals. For instance, measurements are taken at four distances: the reference point, 500 meters, 1000 meters, and 3 kilometers, with an average received power of 0 dBm. If the model designer needs to measure distances up to 3 kilometers, more than 20 measurements would be necessary.

The Mean Square Error (MSE) in an urban scenario is determined by plotting the path loss exponent with respect to n, which is Equation 2.4. This indicates that the area is not densely populated and has few obstructions like foliage and scattering. The minimum error depends on the total number of data points taken and the difficulty of fitting the curve. The *SD* shows the deviation from mean value. In-house data measurements for ultra-wideband frequencies from 3.1 to 10.6 GHz are conducted in the frequency domain, finding the transfer function first and then taking the inverse Fourier transform to get the impulse response of the channel. Averaging is important because there are many fluctuations that take place.

2.4.8. Log Normal Distribution Model

The log normal distribution is a measure of the received signal level in dB at any given point in the coverage area. It is determined by a distance and path loss exponent, which can be used to model path loss and develop cell site planning systems. This information is crucial for understanding any propagation model.

Equation 2.36, presents a log normal distribution that has a PDF given by $p(M)$ and shown in Figure. 2.10.

$$p(M) = \frac{1}{\sqrt{2\pi\sigma}} e^{\frac{-(M-\underline{M})^2}{2\sigma^2}} \qquad (2.36)$$

Here, M is received (true) signal level. $\underline{M}$ Presents the average value of the signal measured over time. Sigma (σ) is the *SD*.

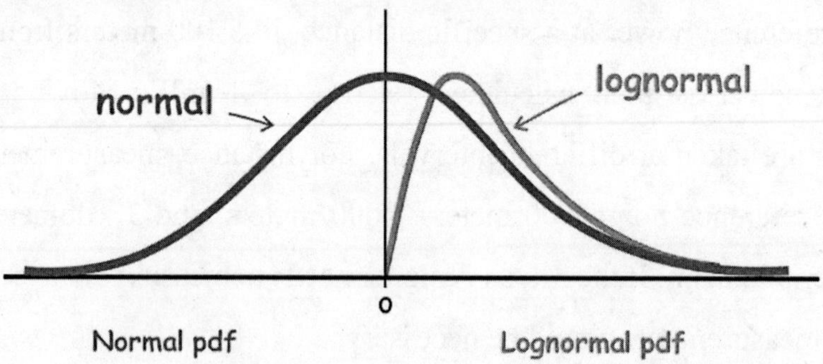

$$f(x) = \frac{1}{\sigma\sqrt{2\pi}} e^{\frac{-(x-\mu)^2}{2\sigma^2}} \qquad f(x) = \frac{1}{\sigma x\sqrt{2\pi}} e^{\frac{-(\ln x-\mu)^2}{2\sigma^2}}$$

Figure 2.10 PDF of received signal level following Log-normal distribution

This Equation 2.37 in log normal shadowing environments can be depicted as follows: $P_r(d)$at a distance "d" are normally distributed in dB about a distance dependent mean with a standard deviation sigma from a transmitter $P_t\ (d)$minus path loss at a distance d denoted by $\underline{P}L\ (d_o)$

$$P_r(d)[dBm] = P_t\ (d)[dBm] - \underline{P}L\ (d_o)[dB] \tag{2.37}$$

$$P_r(d)[dBm] = P_t\ (d)[dBm] - \underline{P}L\ (d_o) + 10n\ log\ \log\left(\frac{d}{d_o}\right) + X_\sigma\ (dB)$$

2.4.9. Percentage of Coverage Area

In a mobile network, computation of coverage area i.e. the ability of a network to deliver signal coverage is an important evaluation factor. Sometimes, due to shadowing effects some areas within a cell will receive signal strength below threshold. This is due to the random scattering effects, which can result in people not receiving enough power. In order to determine the percentage of coverage area we can assume coverage area having simplest circular geometry (*R* being radius) from a BS. Further, minimum threshold level can be defined as gamma. The percentage of the cell is given by PA gamma, which is a function of gamma. If gamma is reduced, more area will be covered, which indicates the probability of received power being greater than threshold (gamma).

2.5. Outdoor Propagation Models

It is obvious that radio propagation outdoors will be over irregular terrain. Outdoor propagation models are designed to predict received power and route loss in the presence of obstructions such as towers, street effects, deserts, and bodies of water, as well as uneven terrain like hills, trees, and buildings. These factors have an impact on path loss and propagation, and different models take these effects into account to establish their outcomes. These models should be as generic as possible and take the topographical profile into consideration. Thus, outdoor radio broadcasts should be able to predict (i) route loss and (ii) received power, independent of the surrounding environmental conditions.

Depending upon geometry, outdoor propagation can be subdivided into three broad categories namely macro cells, microcells, and street microcells. Macro cells are base stations located at high points on towers, buildings, or tall buildings and spanned over several kilometers. Generally, path loss is computed in dB having a normal distribution, reception of a signal at a point is a result of scattering over a large number of obstacles, each of them contributing in a random fashion with a multiplicative factor. Compared to microcells, macro cells have relatively mild propagation characteristics, small multipath delay spread, and shallow fading, implying the feasibility of transmission with larger data rate. As cells grow into smaller cells, their data sending capacity increases, making them mostly used in crowded urban environments. In macro cells, base stations must be as high as possible to increase the probability of line of sight to the receiver and less blockage.

However, there is an interesting scenario when the base station height is not very high, almost at the level of the receiver. This is because increasing the height may cause interference to other base stations and cells where the frequency is being reused. Therefore, the base station height should be only as high as required but not any higher because you have a chance of sending out more interference signals. In street microcells, propagation of signal power will be mostly along the streets. An analogy is made for indoor areas with long corridors and hallways. The path loss exponent "n" can be less than 2 inside buildings where there is a LOS and corridors. The signal may reach with LOS path if the receiver is along the same street or via an indirect propagation mechanism if the receiver turns to another street. A bird's eye view of a densely populated urban area with many buildings can be used to illustrate this concept is shown in Figure 2.11.

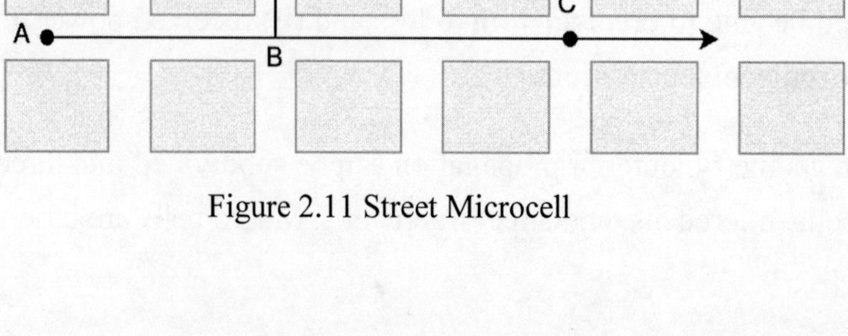

Figure 2.11 Street Microcell

A transmitter at A can easily access either A or C due to the great guiding effect of the streets. If B moves to D, the signal must propagate along this line, diffracting, scattering, or reflecting before going through the street guiding effect and reaching D.

This type of propagation is useful when planning base station locations for dense urban environments, as it takes into account street effects and the height of the base station. The measurement data will also differ in the scenario.

Table 2.1 shows a tabular summary of comparison of Macro-cell Vs. Microcell in terms of various features.

Table 2.1: Comparison of Macro Cell vs. Microcell

Features	Macrocell	Microcell
Cell Radius	1 to 20 km	0.1 to 1 km
Tx power	1 to 10 W	0.1 to 1 W
Fading	Rayleigh	Nakagami Rice
RMS Delay Spread	0.1 to 10 μs	10 to 100 ns
Maximum Bit Rate	0.3 Mbps1	1 Mbps

2.5.1. Longley Rice Model

The Longley Rice Model serves as a tool for point-to-point communication across a broad spectrum ranging from frequency band starting from 40 MHz up to 100 GHz, accommodating different terrains. This model utilizes terrain path geometry and tropospheric refractivity for calculations, factoring in parameters like transmission frequency, length of path, polarizations, heights, reflectivity of terrain, conductivity, dielectric constant, and climatic conditions etc. Often presented as an interactive computer program, it considers several variables for accurate assessments. However, the Longley Rice Model overlooks buildings and foliage, assuming that tower antennas on the top structures are unaffected by trees or buildings. In today's wireless communication environments, these elements are crucial, rendering the model inadequate for mobile communication. In essence, the Longley Rice Model is valuable for predicting path loss in point-to-point communication by considering factors like terrain and geometric optics. Nonetheless, for contemporary wireless communication needs, it falls short as it fails to address potential interference from buildings and foliage on transmission tower antennas.

2.5.2. Okumura Model

The Okumura Model, introduced by Japanese enthusiast Okumura, serves as a popular tool for signal prediction in urban settings. Initially rooted in empirical research, Okumura conducted a thorough measurement campaign in 1968, leading to a model solely based on these measurements. This model, known for its reasonable accuracy, remains widely utilized by communication design companies in Japan. Okumura identified a simple power law as a reliable model for path loss, a finding supported by extensive simulations. The exponent "n" varies based on factors like frequency, antenna heights, and terrain. The path loss exponent is crucial, and detailed measurements affirmed a straightforward power relationship between received power and transmitter-receiver separation.

The Okumura model is suitable for frequency bands from 150 MHz to 1.9 GHz, with extrapolation potential up to 3 GHz. The model's curves are smooth and aid in predicting path loss values.

Okumura created a series of curves depicting median attenuation relative to free space in urban areas with relatively flat terrain. These results are applicable across the frequency spectrum of 150 MHz to nearly 2 GHz, extendable up to 3 GHz, and valid for path length between 1 km and 100 km.

From a mathematical standpoint, the Okumura model can be represented as follows:

$$L_{50}(d)[dB] = L_F(d) + A_{mu}(f, d) - G(h_{te}) - G(h_{re}) - G_{area} \tag{2.38}$$

Here, L_{50} is the 50th percentile (median) value of propagation path loss, $L_F(d)$ is the free space propagation loss, $A_{mu}$and is the median attenuation relative to free space, $G(h_{te})$ the antenna height gain factor of base station, $G(h_{re})$ is the antenna height gain factor of mobile, $G_{\ area}$ is the environmental gain.

The Okumura curves are a set of curves that assess the median attenuation across various carrier frequencies. These curves are utilized to examine path loss discrepancies in urban, semi-urban, and rural settings, offering a straightforward and fairly precise analysis.

It relies on the operational frequency and can be extrapolated up to 3 GHz, despite the curves being available only up to 1.9 GHz. While the Okumura model is highly reliable for suburban and urban regions, its accuracy diminishes in sub-urban areas. Notably, the correction factor decreases at specific frequencies, like 100 MHz, and the model's performance weakens in quasi open spaces, whereas it is crucial in open areas.

For instance, when calculating median loss relative to free space loss at a 50 km distance with a BS height equal to 200 m, MS height of 3 m, and a frequency of 1 GHz, the aim is to compute the average path loss under those conditions. The curves are instrumental in determining median attenuation concerning free space, with the x-axis indicating frequency in MHz and the y-axis representing median attenuation in dB. The median attenuation is around 45 dB relative to free space, serving as a benchmark for median attenuation. Therefore, the Okumura curves present a valuable tool for assessing path loss variations in different environments, offering simplicity and accuracy across various scenarios.

2.5.3. Hata Model

Hata proposed an empirical formula for Okumura's graphical path loss data, applicable within the frequency band from 150 MHz up to 1.5 GHz. The Hata model is an extension of the Okumura model, which was established in urban settings. The median path loss is represented by:

$$L(dB) = 69.55 + 26.16 \times log\ log\ f_c\ (MHz) - 13.82 \times log\ log\ h_{te}\ - a\big(h_{re}(m)\big) + [44.9 - 6.55 \times log\ log\ h_{te}\ (m) \times log\ log\ d\ (km) \quad (2.39)$$

Where,

f_c = Frequency in MHz

h_{te} = BS antenna height (30m to 200m)

h_{re} = mobile antenna height (1m to 10m)

d = T-R separation in km

$a(h_{re}(m))$= is a correction factor

For these values to hold true, the carrier frequency should be in MHz, the distance in km, and the heights of the receiver and transmitter antennas in meters. The transmitter antenna height is a significant factor in the logarithm of the distance, exerting a considerable influence. The Hata model's accuracy ranges from 30 m to 200 m, with MS heights varying from 1 m to 10 m, encompassing a wide range of scenarios. For outdoor propagation path loss computation it should be noted that in urban areas of large cities it will be the maximum. In urban areas of medium and small cities clearly the path loss will go down compared to large cities. In suburban areas it will go down further and in open areas, path loss goes down close to free space propagation. The path loss can be exactly predicted by the Hata model.

In this unit, till this point we have discussed outdoor propagation. On the other hand, a lot of wireless communication takes place indoors via wireless local area network, personal area networks, body area network etc. Next section will discuss indoor propagation models.

2.6. Indoor propagation models

Indoor propagation models differ from outdoor counterparts in two main ways namely (a) in indoor models distance measurement variations are much smaller and

(b) the environment variability is larger for a small transmitter to receiver separation.

(c) The room environment changes more frequently, making indoor channels time-varying.

(d) Even the movement of people inside the room can change the channel considerably and the propagation inside a building is affected by the layout of the building, construction materials, and building type.

Indoor propagation is also characterized by the same mechanisms as outdoors- reflection, scattering, and diffraction. However, entities that affect propagation are much more different, such as doors, windows, partitions, and floor levels. Indoor channels are classified as either LOS or obstructed with clutters. Various types of buildings have different characteristics, and researchers have carried out extensive measurements to come up with models. Computer models can process inputs like type of building, partitions, walls, concrete, space, and room layout to predict indoor propagation characteristics.

Furthermore, indoor propagation causes temporal fading for both stationary and mobile terminals because of environmental changes brought on by workers moving machinery, opening and closing windows, and other events. When there is an impediment, portable receivers get affected as Rayleigh fading; for stationary receivers, Rician fading happens when there is no clear line of sight. Multipath spread is impacted when a signal travels along several routes obtained by diffraction, reflection, or scattering before reaching the receiver.

Root mean square (RMS) delay spread, which is often small (between 30 and 60 ns), is characteristic of buildings with rigid partitions and fewer metals. Inter-symbol interference may arise if the delay spread widens. Equalization is therefore required to counteract the consequences of multipath spread. However, larger buildings can have larger RMS delay spreads, even as large as 300 ns, which degrade performance at high data rates. Therefore, delay spread within the room environment has a direct impact on the data rate. When designing wireless local area networks, such as IEEE 802.11 b, it is essential to consider delay spread aspects.

While defining the path loss, n and σ depend on the type of building and as the value of σ decreases the accuracy of the path loss model increases. The path loss is given by:

$$PL(d)[dBm] = PL(d_0) + 10 \times n \times log\,(dId_0) + X_\sigma \qquad (2.40)$$

Path loss within a building can be classified as hard partitions or small partitions. Hard partitions, like cardboards or plywood, can be fixed or movable, dividing rooms. The path loss for partitions is distinct from that between floors, with specific models proposed based solely on partitions. Computer software can reasonably forecast received signal strength based on partition frequency.

Partition losses between floors are influenced by the building's external dimensions and materials. For instance, receiving a signal from an access point on the first floor on the second floor can result in varying losses, influenced by construction materials, external environment, window count, and window tinting, causing specific types of attenuation.

In a three-floor building, predicting received signal power on different floors is beneficial. Home cell planning assists in determining coverage and access point placement for wireless local area networks. Average values for partition loss between floors are expressed as the Floor Attenuation Factor (FAF) in db. For instance, passing through one floor results in around 13 dB loss, two floors about 18.7 dB, three floors 24 dB, and four floors approximately 27 db. These values aid in determining coverage and access point placement for providing wireless local area networks across the campus. These measurements were conducted at close to 900 MHz, and the values change with higher frequencies, such as 2.4 GHz in the ISM band.

2.6.1. Noise Figure for Link Budget

Correct estimations for receiver noise and Signal-to-Noise Ratio (SNR) at a receiver end is crucial for assessing coverage and service quality standards in mobile communication systems. SNR plays a vital role in determining link quality and error probability in mobile communication systems. Therefore, accurately estimating SNR is essential for establishing connections under different propagation conditions.

In earlier sections, we have already seen computation of received power using path loss equations. However, that may additionally require determining the noise. Computing noise at the receiver input requires knowledge of the gains (or losses) of each receiver stage plus noise temperature at the receiver terminals. Electronic components in working conditions produce thermal noise, contributing to the overall noise level that is observed at the detector terminal of a receiver. To standardize the noise at the output of a receiver the concept of noise figure, F, is employed, which is defined as:

$$F = \frac{Measured\ noise\ power\ out\ of\ device\ at\ room\ temperature}{Power\ out\ of\ device\ if\ device\ were\ noiseless} \quad (2.41)$$

In general, $F > 1$. Further, relation of noise figure (F) to the effective noise temperature (T_e) of a device is given by relation,

$$T_e = (F - 1)T_0 \quad (2.42)$$

Where, T_0 is ambient room temperature (typically taken as 17°C to 27°C). Noise temperatures are often measured in Kelvin, where $0\ K$ is absolute zero or -273°C.

Note that antenna temperatures do not always match physical temperatures. For instance, satellite dish antennas beaming towards space have low temperatures (e.g., $50\ K$) due to space being cold. In contrast, antennas beaming towards Earth have temperatures corresponding to Earth temperature. Consider a passive component with resistive load working at room temperature. It transfers a noise power that is computed as,

$$P_n = kT_0B \quad (2.43)$$

Where, k is Boltzmann's constant given by $13.8 \times 10^{-23}\ Joules/K$ and B is the bandwidth of the measuring device. For passive devices (e.g. transmission lines or attenuators) device loss (L in dB) is equal to the noise figure of the device. That is,

$$F = L \quad (2.44)$$

Antennas with passive elements are treated as having unity gain, regardless of their radiation pattern's measured gain. Computation of noise figure and noise temperature are beneficial in communication analysis, as they allow for the assessment of overall noise amplification without the need to consider the gains of the receiver stages. Consider that at room temperature a resistive load is connected to the input of the receiver. Further, considering receiver has a noise figure F, the noise power at the output of the receiver, referenced to the input, is

$$P_{out/Ref.in} = FkT_oB = \left(1 + \frac{T_e}{T_o}\right) kT_oB \quad (2.45)$$

Actual noise power output (P_{out}) is computed as

$$P_{out} = G_{sys}FkT_oB = \left(1 + \frac{T_e}{T_o}\right) G_{sys}kT_oB \quad (2.46)$$

Where, G_{sys} is the overall receiver gain.

For a cascaded system, the noise figure of the overall system may be computed as,

$$F_{sys} = F_1 + \frac{F_2 - 1}{G_1} + \frac{F_3 - 1}{G_1 G_2} + \ldots \quad (2.47)$$

Or equivalently,

$$T_{sys} = T_1 + \frac{T_2 - 1}{G_1} + \frac{T_3 - 1}{G_1 G_2} + \ldots \quad (2.48)$$

Where, gains are in absolute (not dB) values.

2.7. Multi Path Fading Shadowing, Fading Margin, Shadowing Margin

Multipath fading is a temporary occurrence resulting from the interference of multiple versions of a transmitted signal reaching the receiver at slightly different times. This phenomenon, more common and short-lived, entails variations in signal amplitudes and phases due to path differences.

In wireless channels, fading happens when the mobile device receives the signal directly from the base station or via reflections from the ground, buildings, or other structures. Urban areas with dense structures may experience more reflections, leading to energy dispersion through scattering or diffraction. Different delays, amplitudes, and phases of the same signal are received due to multiple reflections.

Multipath fading is often depicted as a multiplicative factor, where the received signal is adjusted by a factor $\alpha e^{j\theta}$ to represent the phase change. It is crucial to understand this phenomenon and employ effective strategies to counteract its effects by comprehending the mechanisms at play.

Multipath fading and fading are typical occurrences in communication that lead to swift fluctuations in strength of signal over short distances or time periods. On multipath signals, Doppler shifts influence both amplitude and carrier frequency variations. Time dispersions stemming from multipath propagation delays cause the appearance of several replicas of a narrow pulse, particularly in ultra-wideband systems that utilize narrow pulses for transmission. It is essential to grasp these phenomena to enhance communication efficiency and mitigate the effects of fading on communication.

In a static setting, fading is minimal due to the fixed point-to-point micro-wavelength between the transmitter and receiver. However, in dense urban areas with obstacles like buildings, vehicles, and towers, fading becomes more noticeable. The signal characteristics change as a mobile device moves, leading to a perceived temporal variation. As the receiver moves through time, either in distance or velocity, the received signal fluctuates. Small-scale fading is influenced by various factors such as scatterers on the ground, buildings, vehicles, and towers, along with multipath components. Understanding the impact of mobility on fading is crucial as the received signal fluctuates when the receiver moves over time.

Fading poses a significant challenge in mobile wireless communications, stemming from multipath propagation, reflective surfaces, and scatterers. These factors lead to multiple signal versions reaching the receiver with varying amplitudes and time delays. The HF channel, functioning as a multipath channel, experiences fading due to ionospheric reflections.

The velocity of the mobile device also impacts small-scale fading, introducing a temporal fading aspect. Higher mobile speeds alter the fading pattern, while slower speeds cause Doppler shifts in each signal multipath component. Additionally, the angle of MS relative to the BS influences the fading pattern. Different multipath signals arriving at various angles undergo distinct Doppler shifts, combining delayed signal copies that differ from the transmitted signal. To mitigate fading effects, addressing random frequency modulation and adjusting data rates is crucial. Typically, the Doppler shift can be +ve or -ve depending on the MS's direction towards or away from the BS.

Small-scale fading is affected by various factors, such as the speed of nearby objects, environmental changes, and multipath components. In wireless local area networks, environmental movement predominantly influences small-scale fading. For instance, in room settings, the movement of people and spinning fans can impact small-scale fading. When surrounding objects move faster than the mobile device, this movement takes precedence over fading effects. In wireless local area networks, any environmental movement becomes the main cause of small-scale fading.

The concept of coherence time, used to assess the channel stability, determines how quickly the channel undergoes changes. In multipath scenarios, coherence time differs between room environments and rural or urban settings. By examining the correlation between coherence time and Doppler shift, we can determine whether channel variations are due to surrounding object movements or independent of MS mobility.

2.8. Statistical Models for Multipath Fading Channels

Numerous multi-path models are proposed to elucidate the statistical characteristics of mobile networks. The initial model, introduced by Ossana, focuses on wave interference from structures like buildings. While Ossana's model accurately predicts flat fading power spectra in suburban regions, it assumes the presence of a direct transmission path between sender and receiver, limiting its applicability to specific reflection angles. This rigid model is unsuitable for urban settings where direct paths are typically obstructed by buildings or other structures. In contrast, Clarke's model based on scattering as main phenomena, is a commonly adopted alternative. Clarke presented a model that uses scattering to infer the statistics based properties of electromagnetic fields in the signal received by a mobile device. The model makes the assumption about mobile antenna placements. It also assumes that the transmitter is stationary and the average amplitude of each wave is the same at the receiver after experiencing comparable attenuation at short distances. In Figure 2.12, there is an illustration of plane waves reaching at MS. It is assumed that MS is moving at a constant velocity v in the +ve x-direction, with the angle of arrival as shown.

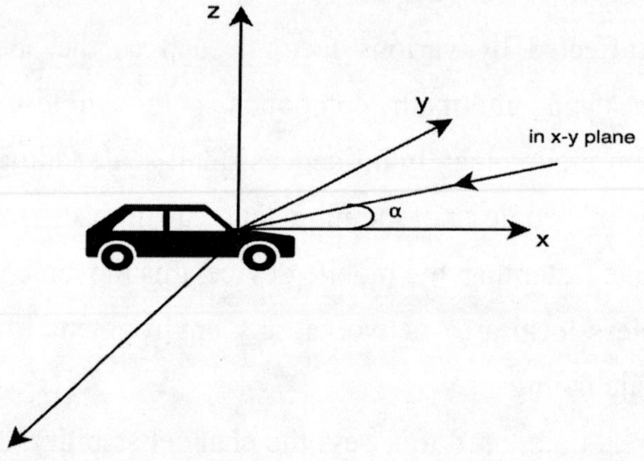

Figure 2. 12 Illustrating plane waves arriving at random angles.

It should be noted that each wave reaching the mobile device encounters a Doppler shift due to the receiver's movement and arrives concurrently at the receiver, without additional delays from multipath (flat fading assumption). For such the nth wave arriving at an angle α_n to the x-axis, the Doppler shift can be computed as:

$$f_n = \frac{v}{\lambda} cos\alpha_n \qquad (2.49)$$

Where, λ- is the wavelength of the incident wave.

It should be noted that as the Doppler shift is very small when compared to the carrier frequency, all electromagnetic components can be approximated as Gaussian random variables if N is sufficiently large and after following computing steps the random received signal envelope r with a Rayleigh distribution given by

$$p(r) = \frac{r}{\sigma^2} exp\left(-\frac{r^2}{2\sigma^2}\right) \qquad 0 \leq r \leq \infty \qquad (2.50(a))$$

$$p(r) = 0 \qquad r < 0 \qquad (2.50(b))$$

Where, $\sigma^2 = \frac{E_0^2}{2}$

UNIT SUMMARY

This module started with thorough knowledge of wireless propagation, covering many areas that play vital role diversity plays in wireless communication systems. It provided knowledge about important topics such as link budget, which covers transmitter power, antenna gains, and losses in case of estimation or computation of signal strength.

The module also examined free-space route loss, illustrating deterioration of signals in open areas with the increase in distances. It discussed about multipath fading, shadowing, fading margin, and shadowing margin of the receiver, opening scope to mitigate signal deterioration. Through this unit, there was a better insight towards understanding of wireless propagation phenomena, will prove essential for design and operation of wireless communication networks.

EXERCISES

Multiple Choice Questions

1. Which of the following describes statistical models for multipath fading channels?

   a) They offer deterministic forecasts for signal strength fluctuations.

b) They use physical concepts to describe signal propagation.

c) They use statistical distributions to model signal variations.

d) They are appropriate for simulating signal propagation in free space.

2. Which statement best represents the noise figure used in link budget calculations?

a) The noise figure depicts the highest power level of noise in a communication channel.

b) The noise figure represents the signal's attenuation during transmission.

c) The noise figure quantifies the decrease of the signal-to-noise ratio caused by the receiver.

d) The noise figure measures the transmitter's efficiency at converting signal power to noise power.

3. Which of the following factors is frequently included in indoor propagation models?

a) Terrain features

b) Weather conditions

c) Construction materials and impediments

d) The distance between transmitter and receiver.

4. Which of the following assertions characterizes shadowing models in wireless communication?

a) Shadowing models are primarily concerned with signal reflections.

b) Shadow models are deterministic in nature.

c) Shadowing models are mostly used in free-space propagation

d) Shadowing models account for obstacle-induced signal attenuation.

5. Which of the following is a feature of mobile wireless propagation?

a) It exclusively addresses signal behavior in stationary settings.

b) It takes into account signal attenuation caused by obstructions in motion.

c) It only uses deterministic models for signal prediction.

d) It does not account for the impacts of multipath fading

Answers to Multiple Choice Questions

1 c), 2 c), 3 c), 4 d), 5 b)

Short and Long Answer Type Questions

Short Questions

1. In what manner does the Free Space Propagation Model depict signal propagation?
2. What are the fundamental assumptions of the Free Space Propagation Model?
3. Which factors influence signal propagation in free space?
4. Provide an explanation of equivalent noise temperature in the context of electronic systems and its relationship with noise figure.
5. Define cascaded systems in terms of communication networks, and explain how noise builds in them.
6. How can man-made noise be reduced in communication systems?
7. Describe how diffraction influences the propagation of electromagnetic waves around barriers.
8. What is the Free Space Propagation Model, and what are its main assumptions?
9. How does the Free Space Propagation Model facilitate the understanding of wireless Communication systems?
10. What factors influence signal transmission based on the Free Space transmission Model?
11. How does multipath propagation affect small-scale fading?
12. What causes large-scale path loss in wireless communication?
13. What are the limits of the Log-Distance Path Loss Model?
14. Describe the Log Normal Shadowing Model and its use in wireless communication.
15. Explain the probability density function for the Log Normal Distribution Model.
16. Explain the idea of path loss in outdoor propagation models and how it affects wireless communication.

Long Questions

1. How is the attenuation of electromagnetic waves in free space mathematically characterized by the Free Space Propagation Model, and what is its significance in wireless communication?
2. What are the underlying assumptions of the Free Space Propagation Model, and how do they simplify the analysis of wireless communication systems?
3. In real-world scenarios, how do environmental factors like atmospheric conditions and obstacles affect the accuracy of the Free Space Propagation Model?
4. Define the concept of equivalent noise temperature within electronic systems and elucidate its correlation with noise figure. Elaborate on the calculation of noise figure and its importance in the design of amplifiers. Discuss the impact of augmenting the equivalent noise temperature on the performance of a communication system.
5. Explore the meaning of equivalent noise temperature in the realm of electronic systems and its connection to noise figure. Detail the methodology behind calculating noise figure and its relevance in the design of amplifiers. Analyze the consequences of elevating the equivalent noise temperature on the efficiency of a communication system.
6. Explain the ground reflection (2-ray) model in wireless communication. How does the ground reflection model influence signal transmission in urban and rural settings?
7. How does scattering affect multipath propagation and signal fading in wireless communication systems?
8. Compare and contrast various small-scale propagation models, including Rayleigh fading and Rician fading.
9. Discuss the Log Normal Distribution's mathematical features and how they apply to real-world propagation situations.
10. Evaluate the usefulness of outdoor propagation models in real-world circumstances, taking into account accuracy, complexity, and processing needs.
11. Explain the fundamental assumptions employed in indoor propagation models and how they affect model accuracy.

12. Define multi-path fading, shadowing, fading margin, and shadowing margin in the context of wireless communication systems. Explain their importance in maintaining dependable communication.

NUMERICAL PROBLEMS

1. Let's consider a scenario where a transmitter is emitting a signal at a frequency of 2.4 GHz (Wi-Fi frequency) in a free space environment. The power of the transmitter is 100 mW, and we aim to determine the power received at a distance of 500 meters from the transmitter. The parameters provided are as follows: Frequency (f): 2.4 GHz (2.4 x 10^9 Hz), Transmitter Power (Pt): 100 mW (0.1 W), Distance (d): 500 meters.

2. Assume we have a transmitter broadcasting a signal with 10 watts of strength at 2.4 GHz (Wi-Fi frequency) to a receiver 1 kilometer away. In this case, the signal travels straight from the transmitter to the receiver as well as through reflection off the ground. Compute the received power at the receiver using the 2-ray model, assuming perfect reflection (no loss).

3. A radio tower that broadcasts a signal at the frequency of 100 MHz. A 50-meter-tall building stands in the way of the skyscraper. Calculate the diffraction loss caused by the blockage.

4. Consider a 3.5 GHz wireless communication system in an urban context. The Log-Distance Path Loss Model is used to forecast path loss, while the Log-Normal Shadowing Model is used to account for shadowing effects. The route loss exponent (n) is 3.8, the reference distance (d0) is 1 meter, and the standard deviation of shadowing (σs) is 5 db.
   a. Use the Log-Distance Path Loss Model to calculate the path loss over a 500-meter distance.
   b. Use the Log-Normal Shadowing Model to calculate shadowing loss at the same distance.
   c. Calculate the overall path loss by taking into account both path and shadowing losses.
   d. Discuss how shadowing affects the reliability of the communication system.

5. Consider a 2.4 GHz wireless communication system in a suburban context. The Okumura and Hata model is used to forecast route loss in the outdoor environment. Use the Okumura

and Hata model to compute the route loss given a 1.5-kilometer distance between the transmitter and receiver.

6. Assume we are creating a wireless communication network with a frequency of 2.4 GHz. The transmitter's output power is 20 dBm, while the receiver's sensitivity is -90 dBm. The link budget must account for the receiver's noise figure, which is provided as 3 decibels.
   a. Calculate SNR at the receiver input while ignoring the noise figure.
   b. Determine the noise power at the receiver's input using the noise figure.
   c. Calculate the new receiver sensitivity, including the noise figure.
   d. Determine whether the communication link will have the necessary signal-to-noise ratio for reliable communication.

PRACTICAL

1. Develop a program that calculates path loss for both indoor and outdoor propagation models, taking into account relevant environmental conditions. Create routines for each model and give the user options to select the desired model. The program should prompt the user for input parameters such as distance, frequency, and environmental conditions, and then display the computed route loss. Make sure that the outdoor propagation model takes into consideration terrain, building materials, and vegetation.
2. Create a program that calculates the noise number for a wireless communication link budget. The program should prompt the user to enter important settings including transmitter power, receiver sensitivity, bandwidth, and ambient conditions. It should then calculate the noise figure using these inputs and output the result.

KNOW MORE

Propagation models for the Internet of Things (IoT) and sensor networks are critical in optimizing the design and deployment of wireless communication systems customized to the specific needs of IoT devices. These models take into consideration the specific characteristics and restrictions of IoT and sensor networks in order to improve energy efficiency, coverage, and connectivity. Here's a quick overview:

a. Energy Efficiency: IoT devices are frequently battery-powered and have limited energy reserves. Propagation models for IoT networks prioritize energy efficiency by optimizing transmission power levels, scheduling transmissions efficiently, and utilizing low-power communication protocols. By precisely forecasting signal propagation and network quality, these models assist IoT devices in conserving energy while ensuring reliable connectivity.

b. Coverage and Connectivity: IoT networks may include a significant number of devices distributed in a variety of settings, including indoor, outdoor, and harsh industrial environments. Propagation models provide for the effects of obstructions, building materials, and environmental variables on signal propagation to ensure adequate coverage and connectivity. These models help determine the best device placement and network design for dependable communication by analysing signal attenuation, path loss, and multipath effects.

c. Path Loss and Fading: Propagation models for IoT networks take into account path loss and fading phenomena including shadowing and multipath propagation. By precisely modelling these effects, the models provide insights into signal strength changes and allow for the optimization of communication protocols and signal processing techniques to reduce the influence of fading on data transmission reliability.

d. Localization and Positioning: IoT applications frequently demand accurate device localization and positioning for asset tracking, indoor navigation, and context-aware services. Propagation models enable localization approaches based on signal strength measurements, time-of-arrival, angle-of-arrival, or fingerprinting. These models make it easier to design precise localization algorithms for IoT and sensor networks by simulating signal propagation.

To summarize, propagation models for IoT and sensor networks take into account the particular requirements of low-power, resource-constrained devices deployed in a variety of contexts. These models make it possible to build, optimize, and deploy efficient and reliable wireless communication systems for IoT applications by offering insights into signal propagation characteristics and connection quality measurements.

Dynamic QR Code for further Reading

FURTHER READING

1. Rappaport, Theodore S. Wireless communications: principles and practice. Cambridge University Press, 2024
2. NPTEL course: https://archive.nptel.ac.in/courses/117/102/117102062/

3 Diversity, Capacity and MIMO Systems

UNIT SPECIFICS

Through this unit, the following aspects are discussed:

- *Diversity, types of diversity and Space diversity techniques*
- *MIMO antenna systems capacity*
- *MIMO-based architecture and capacity*

RATIONALE

This module offers a comprehensive examination of the imperative role diversity plays in modern wireless communication systems. Beginning with an exploration of the compelling need for diversity, students will gain insight into the fundamental reasons behind its incorporation in communication infrastructure. Delving into the various types of diversity, including spatial, frequency, time, and polarization diversity, students will develop a nuanced understanding of the multifaceted approaches to enhancing signal reliability and performance. A significant focus of the module is directed towards space diversity techniques, encompassing Selection Combining, Maximum Ratio Combining, Equal Gain Combining, Square Law Combining, and Minimum Input Minimum Output (MIMO) antenna systems.

Through detailed explanations and illustrative examples, students will grasp the operational principles and applications of each technique, thus cultivating proficiency in selecting and implementing appropriate diversity schemes. Moreover, this module addresses the pivotal aspect of antenna systems' capacity within the context of space diversity. By elucidating concepts such as antenna gain, radiation patterns, and spatial multiplexing, students will acquire the foundational knowledge essential for optimizing system capacity and spectral efficiency. By the conclusion of this module, students will emerge equipped with a comprehensive understanding of the indispensable role of diversity in wireless communication systems. They will possess the requisite knowledge to design, analyze, and optimize robust communication systems capable of overcoming the challenges posed by fading channels, interference, and spectral limitations.

PRE-REQUISITES

- Mathematics (Class XII)
- Basics of Signal and Wave
- Basics of Communication system

UNIT OUTCOMES

After going through this unit students will be able to:

U3-O1: Describe Diversity, its advantages and types of diversity

U3-O2: Analyze system capacity due to various space diversity techniques used in practice

U3-O3: Describe Smart antenna and MIMO including architecture and MIMO channel capacity

Unit-3 Outcomes	***MAPPING of UNIT OUTCOMES WITH COURSE OUTCOMES (1- Weak Correlation; 2- Medium correlation; 3- Strong Correlation)***				
	CO-1	***CO-2***	***CO-3***	***CO-4***	***CO-5***
U3-O1	1	3	1	3	1
U3-O2	2	3	3	2	--
U3-O3	1	2	2	3	2

3.1 Introduction to Equalization and Diversity

Equalization plays a vital role in mobile communications by addressing the problem of Inter-Symbol Interference (ISI) in multipath settings, where signals arrive via multiple paths. It works as an inverse filter by compensating for the anticipation of signal properties like amplitude and delay. However, adaptive equalizers are necessary to tackle channel variations over time due to the unknown nature of the channel.

Employing diversity through multiple receiving antennas at either the transmitter or receiver helps reduce fading effects in various scenarios. These techniques can be applied at Base Station (BS) and Mobile Station (MS), with any diversity technique being suitable for either location. One should observe that equalization techniques are deployed to mitigate ISI effects while; diversity helps to reduce the fading properties (signal's attenuation and fading duration) perceived by a receiver. Equalization has two separate modes, the first is the training mode and the second is the tracking mode. During training mode, the weight factors of the equalizer are set by sending a known signal and then only it can be utilized. Whereas diversity does not have training or requirements related to issues like equalization.

3.1.1 RMS Delay Spread

Root Mean Square (RMS) delay spread specifies the time dispersiveness of the channel observed from the power delay profile of the channel. It indicates a time interval during which received signal strength is above a certain threshold. It is related to the second central moment (r) (spread from the mean) of the power delay profile as:

$$\sigma_\tau = \sqrt{r^2 - \left(\underline{r}\right)^2} \tag{3.1}$$

3.1.2 Coherence Bandwidth (B_C) and Coherence Time (T_C)

Similar to what delay spread does in the time domain, coherence bandwidth (B_C) helps to analyze the channel in the frequency domain. Coherence bandwidth (B_C) is defined as the range of frequency over which the frequency correlation remains greater than the threshold.

Delay spread has its natural existence due to multipath phenomenon while coherence bandwidth is derived out of delay spread. In General, The RMS delay spread (σ_τ) and coherence bandwidth (B_C) are related in inverse proportion to each other. However, the exact relationship between them is a function of the existing multipath environment. If the frequency correlation is greater than 0.9 then B_C can be estimated as,

$$B_C = \frac{1}{50\sigma_\tau} \quad (3.2)$$

And if the frequency correlation is greater than 0.5 then B_C can be estimated as,

$$B_c = \frac{1}{5\sigma_\tau} \quad (3.3)$$

Where, σ_τ-RMS delay spread. σ_τ is directly obtained from the measurement of data and then Equations (3.2) and (3.3) help to compute the coherence bandwidth. It is worth noting that compared to the signal bandwidth (B), larger or smaller values of coherence bandwidth (B_c) help in the identification of the channel type i.e. whether it is frequency selective or frequency flat.

Coherence time (T_c) is a statistical measure of how long a channel's impulse response remains constant across time. It represents how quickly the channel changes over time. If the symbol period (reciprocal of message signal bandwidth) exceeds the channel's coherence time, the channel will tend to change itself in between transmission of signal and distortion develops at the receiver. It aids in determining if a channel is rapid fading or slow fading, allowing for the selection of equalization diversity, coding, or a combination. Coherence time is inversely proportional to the maximum Doppler spread and is mathematically defined as:

$$T_c = \frac{1}{f_m} \quad (3.4)$$

Where, fm is the maximum Doppler spread.

Types of small-scale fading based on Multipath delay Spread is summarized in Figure 3.1 (a) while, small-scale fading based on Doppler spread is presented in Figure 3.1 (b). Small-scale fading resulting from multipath time delay spreading can be categorized into two main types namely (a) flat fading and (b) frequency-selective fading as shown in Figure.3.1 (b). Flat fading arises from variations in the multipath channel's gain, causing fluctuations in the received signal's amplitude over time. Whenever the signal bandwidth (B_S) is much smaller than the coherence bandwidth (B_C); flat fading is observed. Rayleigh fading, a form of flat fading, occurs when the symbol period of the transmitted signal exceeds the channel's delay spread, resulting in deep fades and necessitating increased transmit power.

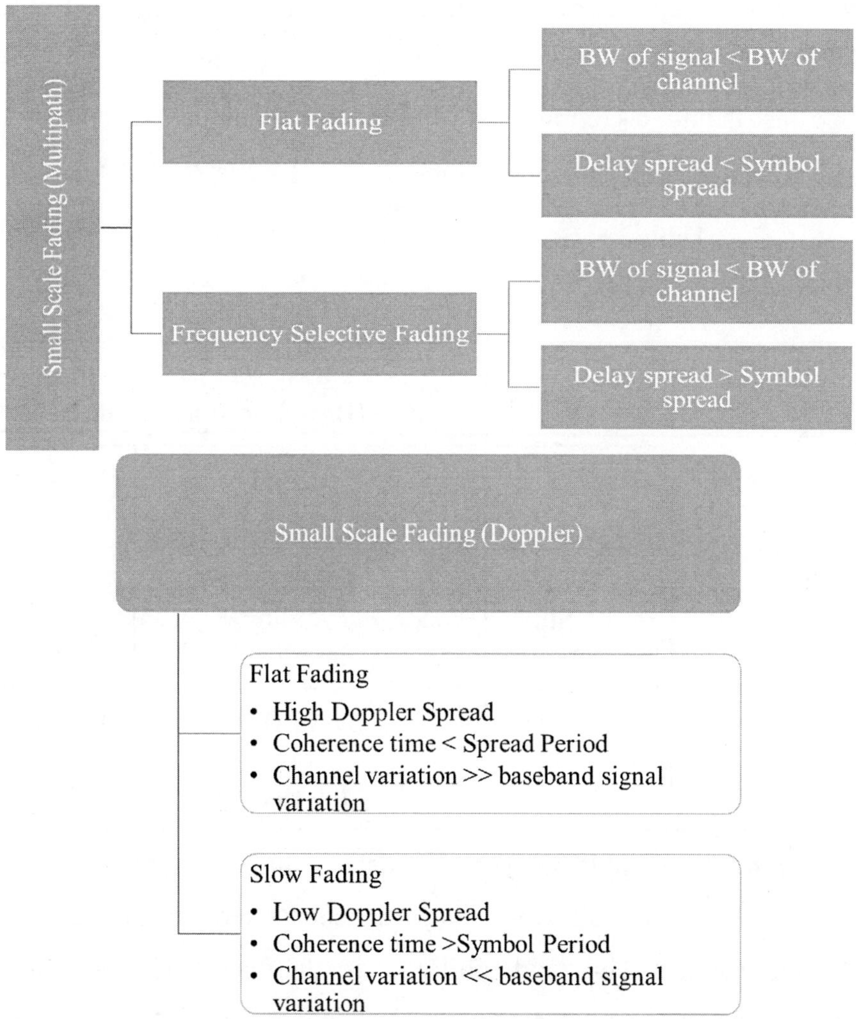

Figure 3.1 (a) Types of small-scale Fading (Based on MultiPath delay Spread) and (b) Types of small-scale Fading (Based on Doppler Spread)

To address these challenges, diversity techniques, such as coding and interleaving, can be implemented. In a frequency-flat slow fading channel, diversity techniques effectively mitigate flat fading issues. Even in instances of fast flat fading channels, diversity strategies remain viable. Coding and interleaving methods also prove beneficial in scenarios characterized by flat fading.

While equalization may not be effective in flat fading scenarios, it becomes indispensable in frequency-selective environments. Equalization serves to counteract channel-induced distortions by acting as an inverse filter. In situations where the channel is both frequency-selective and fast fading, a combination of diversity techniques, coding, and interleaving may be more suitable than equalization alone.

Table 3.1: Countermeasures for Small-scale fading

Fading Countermeasures			
Small Scale Fading (Based on multi path Doppler Spread)		**Small Scale Fading (Based on multi path Delay Spread)**	
		Flat	**Frequency selective**
	Slow	DIVERSITY Coding + Interleaving	EQUILIZATION
	Fast	DIVERSITY Coding + Interleaving	DIVERSITY Coding + Interleaving

A practical guide is available to assist in determining the appropriate techniques or combinations thereof for flat slow, fast flat, frequency-selective slow, and fast frequency-selective fading channels. This guide outlines that the distinction between frequency-flat fading and frequency-selective fading is contingent upon the channel's coherence bandwidth, while the effectiveness of speed fading and interleaving techniques relies on the channel's coherence time. Fading countermeasures are depicted in Table 3.1. On the X-axis, there are frequency flat fading and frequency selective fading. These two depend on the coherence bandwidth of the channel. On the Y- axis there are slow fading and fast fading; they depend on the coherence time of the channel.

In the first location in this grid is a frequency flat slow fading channel where diversity can be used as a countermeasure. Diversity works well for flat fading. Similarly, even if it is flat but fast fading, still the choice of diversity technique is available. On top of that, you can also use coding and interleaving. Interleaving is a technique that causes a little bit of delay in the signal, whereby there is a memory where in read in row wise read out column-wise, at the receiver read in column wise and read out row-wise. Interleaving can be used to spread out the burst errors so both diversity techniques and coding plus interleaving techniques will work for flat fading scenarios. Equalization as a countermeasure for fading becomes effective in a frequency-selective environment. Equalization has been designed to overcome the distortions put in by the channel and is an inverse filter. However, if there is frequency selective but fast fading then there is no time to effectively track the channel and therefore again we may have to go back to diversity and coding and interleaving.

3.1.3 Equalization and Adaptive Equalization

Equalization serves as a pivotal technique employed to mitigate or eliminate the adverse impacts of Inter-Symbol Interference (ISI), a significant impediment to high-speed data transmission. Data transmission rates without equalization are notably limited, underscoring the necessity of equalization for achieving efficient high-speed data transmission. While even GSM incorporates equalization, the standard refrains from specifying a particular type, granting manufacturers the flexibility to employ proprietary methods. Adaptive equalizers monitor the dynamic attributes of the mobile channel, whereas blind equalizers make use of constant amplitude to predict and counteract channel-induced distortion. By leveraging a training sequence, adaptive equalizers enable manufacturers to implement custom equalization approaches. Despite the absence of a mandated type, the importance of equalization in mitigating inter-symbol interference within mobile fading channels remains indisputable.

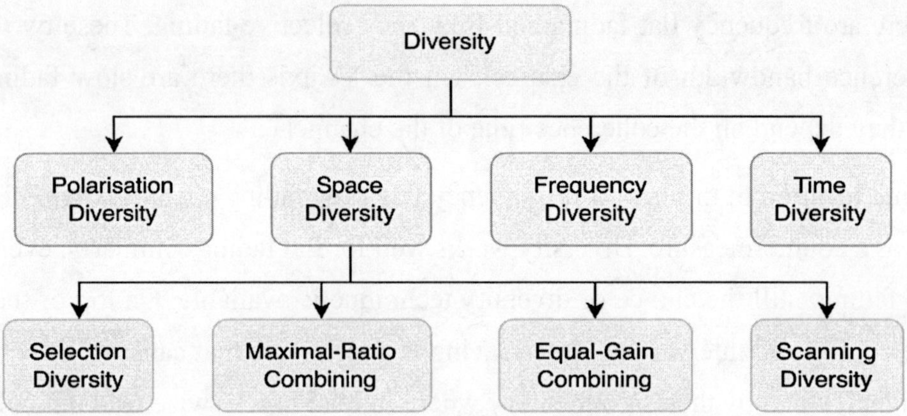

Figure 3.2 Types of Diversity

3.2 Types of Diversity

Diversity serves as a key method to combat fading effects that can degrade modulation performance significantly. Various diversity techniques exist, such as antenna diversity, frequency diversity, time diversity, polarization diversity, angle diversity, and code diversity. This is illustrated in Figure 3.2.

Polarization diversity allows signals of horizontal and vertical polarization to fade independently. Polarization diversity is a method that leverages the decorrelation of two receiving ports to achieve diversity gain. To achieve effective diversity, the two receiving ports must remain cross-polarized, and a correlation coefficient below 0.7 is ideal.

Spacing between antennas at a BS is termed as **space diversity** and it is the most widely used type of diversity. Spacing up to twenty wavelengths is required to be maintained. Conversely, mobile stations can utilize a few wavelengths or even half a wavelength for a less congested environment. The key point to be observed in spatial diversity is the determination of separation between antenna elements by the amount of clattering in the multipath environment. Implementing space diversity at a BS requires significant antenna separation, which may be impractical. Incidentally, polarization diversity at a BS eliminates the need for antenna spacing and only requires two elements: one vertically polarized and the other horizontally polarized. Space diversity at a BS is comparatively less feasible due to the narrow incident field angles and the higher costs involved. In the beginning periods of cellular radio, subscriber units used vertical whip antennas mounted on vehicles. Today, subscriber units are portable phones that have generated interest in polarization diversity at BSs. A theoretical model for polarization diversity assumes a signal sent from an MS with either vertical or horizontal polarization and at BS signals are received by a polarization diversity antenna with two branches. However, the horizontal and vertical polarization paths between MS and BS are somewhat correlated and signal decorrelation in each polarization results from multiple reflections in the channel, leading to signals with varying amplitude and phase reflections. In practice, there is some reliance on received polarization on transmitted polarization.

Frequency diversity involves transmitting and receiving the same information signal simultaneously on multiple independent fading carrier frequencies. This is because frequencies spaced further apart than the channel's coherence bandwidth will not experience the same level of fading. The likelihood of simultaneous fading is determined by the combination of individual fading probabilities. Microwave line-of-sight links often utilize frequency diversity, employing multiple channels in an FDM mode. This method necessitates additional bandwidth and potentially a very large number of receivers to achieve frequency diversity, each set to different frequencies. While this approach may incur added costs, it can be justified for critical communication needs. Despite the bandwidth requirements, frequency diversity can be applied in broad-spectrum scenarios when necessary.

Time diversity is where the transmissions that are longer than the channel's coherence time are separated. This method sends messages with the same content at different times over many channels. Delivering data at intervals longer than the channel's coherence time—which is correlated with the biggest Doppler shift—makes this strategy efficient. Understanding the channel's Doppler shift makes it easier to calculate the coherence time, which establishes the intervals needed for time diversity. To provide a variety of fading conditions, several signal repetitions are required, which enhances diversity. Spread spectrum CDMA with rake receivers is the method used in modern temporal diversity systems to offer message redundancy during transmission.

Antenna diversity involves using multiple antennas separated by the channel's coherence bandwidth.

Code diversity utilizes different codes for achieving code diversity.

Macroscopic diversity combats large-scale fading and shadowing effects. Large-scale fading results from shadowing caused by terrain and environmental variations. To mitigate this, it is crucial to deploy unshadowed antennas, ensuring that not all antennas are within shadowed areas to enhance the signal-to-noise ratio. This might necessitate separating antennas by several kilometers.

On the other hand, **microscopic diversity** addresses small-scale fading, which is more severe due to Rayleigh fading. Small-scale fading arises from multiple reflections from the environment, leading to significant and rapid amplitude fluctuations. To counteract this fading, antennas with robust signals are chosen, and placed at appropriate distances.

3.3 Space diversity techniques

Spatial diversity stands out as one of the most commonly employed techniques, enabling the utilization of multiple antennas either at the BS or across various bases. The crucial aspect of spatial diversity lies in the spacing between antenna elements, which can be adjusted based on the multipath surroundings.

Space diversity, also referred to as spatial diversity, is a technique used in transmission or reception to mitigate fading effects by utilizing two or more physically separated antennas, ideally spaced by at least half a wavelength. With this method, separate and uncorrelated fading pathways can be established at both communication endpoints. Numerous factors, including wavelength, frequency, and the surrounding environment, affect the distance between antennas. A separation of $\lambda/2$ works well in highly cluttered situations with many reflections, but a larger separation works well in less cluttered conditions.

A mobile device usually needs a $\lambda/2$ separation in situations with lots of reflections, but several wavelength separations may be needed at the base station with fewer reflectors to receive diversity. The wavelength and surrounding clutter near the receiver system also impact the spacing between antenna elements.

A general block diagram of space diversity (Figure 3.3) illustrates that signals arriving from spatially distinct antennas on MS exhibit uncorrelated amplitudes for antenna separations of half a wavelength or more. The diagram depicts antenna elements 1, 2, and others, featuring variable gains and logic to select the appropriate gain. In cluttered environments, the output should exceed $\lambda/2$.

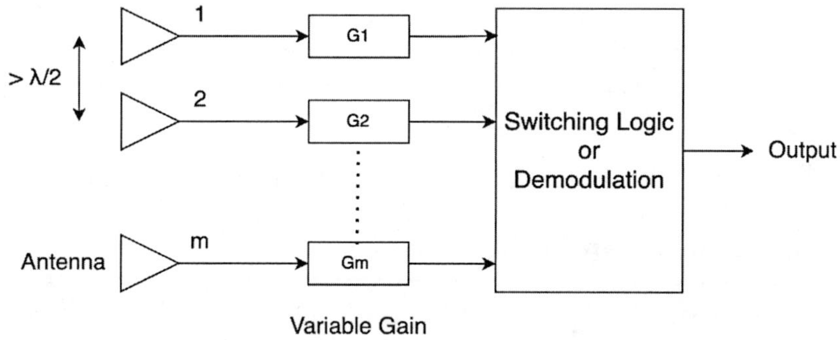

Figure 3.3 Block diagram of space diversity

Diversity can be combined using four techniques: selection, maximum ratio, equal gain, and square law.

3.3.1 Selection Combining (SC)

Selection diversity, a key principle, involves selecting the strongest signal from various branches at the receiving end. The aim is to identify the highest received signal values in receiver systems equipped with multiple antennas. In a block diagram scenario, M-independent Rayleigh fading channels are examined, featuring a transmitter and M fading paths. A mechanism is employed to detect the strongest signal among the branches by opting for the branch with the most robust received signal path. The assumption is that each branch maintains identical average signal strength to estimate the effective Signal to Noise Ratio

(SNR) and probability of error. The selection-combining technique is simple to implement. However, it is not optimal in terms of performance as it overlooks the information provided by all diversity branches except for the one generating the highest power for its demodulated signal.

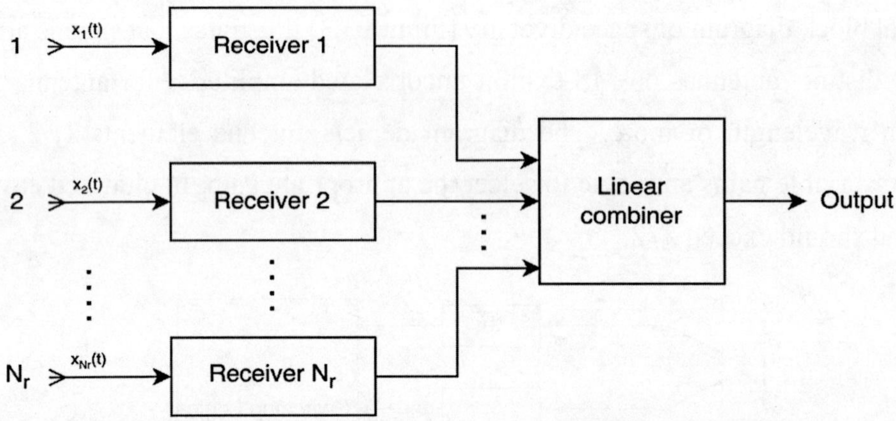

Figure 3.4 Block diagram of MRC using multiple receive antennas

3.3.2 Maximal-Ratio Combining (MRC)

The drawback of a selection combiner technique is addressed by the MRC, as illustrated in the block diagram in Figure 3.4.

The maximal-ratio combiner comprises Nr linear receivers, succeeded by a linear combiner. By utilizing the complex received signal at the kth diversity branch, we determine the complex output from the linear combiner to find out the maximum SNR at each instant of time.

From a theoretical perspective, MRC is considered the most efficient among linear diversity-combining methods, as it delivers the highest possible instantaneous output signal-to-noise ratio.

However, practically there are three essential considerations to bear in mind:

a. Significantly intricate instrumentation is required to fine-tune the complex weighting parameters of the MRC to their precise values.

b. The incremental enhancement in output SNR achieved by the MRC compared to the selection-combiner is relatively modest. Any additional improvement in receiver performance is probably negated by the challenge of accurately configuring the MRC.
c. While a linear combiner prioritizes the selection of a branch with the strongest incoming signal, minor enhancements in overall receiver performance can be achieved through other combiner specifics.

3.3.3 Equal-Gain Combining (EGC)

The discussion leads to the introduction of the EGC, where unlike the branch components values in the MRC, the magnitudes in the EGC are uniformly set to a constant value – unity, for ease of implementation.

MRC, as well as EGC, depend on accurately estimating the phase of various diversity branches and merging signals cohesively. However, this process can be impractical at times due to the physical separation of diversity receivers or hardware constraints. In such scenarios, square-law combining provides an opportunity to leverage diversity advantages without the need for phase estimation.

EGC is a space diversity method that coherently aligns all signals with equal gain for all signal levels to achieve the highest SNR at the receiver consistently. EGC is the preferred choice when variable weighting capability is not feasible. Each branch's gain is normalized to one, enabling EGC diversity. This approach allows the receiver to utilize signals received on each branch simultaneously, ensuring a satisfactory signal output from multiple unfavorable inputs. EGC is slightly less effective than maximal ratio combining but outperforms selection diversity.

These combining methods differ from multipath fading techniques as they focus on overcoming fading effects by concentrating on uncorrelated fading. Multipath signals arrive at each antenna element independently and combine at each element; thus, there is no attempt to isolate the multipath. Although multipath combining can lead to deep signal fades in some situations, spatial diversity techniques target this issue specifically. Rake receivers segregate different multipath components, while time diversity techniques and rake receivers will be elaborated on later in this discussion.

3.3.4 Square-Law Combining (SLC)

In contrast to maximal-ratio combining, SLC is specifically suited for certain modulation techniques. It is most effective with orthogonal modulation techniques, such as Frequency Shift Keying (FSK) or Direct-Sequence (DS) CDMA signals, where distinct (almost orthogonal) frequencies or sequences are utilized to code and transmit different data symbols.

SLC is a wireless communication technique that enhances signal reception by addressing signal fading. It uses squares to boost signal strength while minimizing noise, as well as summarizing squared signals from all antennas. This method tries to increase the SNR and eliminate fading effects, resulting in a more stable and dependable signal. SLC is computationally simpler than other diversity-combining approaches, making it ideal for real-time applications. It is also simple to integrate in receiver designs. However, because it does not require phase information understanding, it may not perform as well as more complex combining approaches and may not be coherent, resulting in performance trade-offs. Overall, SLC has various benefits in wireless communication, including minimal complexity and ease.

3.4 Wireless Channel Capacity

As demand for wireless communication is still growing across the world, it is motivating to study one of the basic measures of performance channel capacity. Capacity can be termed as the maximum rate of communication that can be achieved with permissible error probability. It should be noted that there is an absence of a commonly acceptable way to define capacity for fading channels that is acceptable in all applications. Shannon introduced pioneer work for capacity in the 1940s where increased data rates were predicted and subsequently, a reduction in error probability was proved using coding without compromising data rate or more bandwidth requirement. In fading channels, the utilization of resources like power, diversity and degrees of freedom allowed designers to set various capacity measures. In literature, the capacity of the Additive White Gaussian Noise (AWGN) channel sets the primary standards and then it paves the way to study the capacity measures of different wireless channels.

Capacity for channels needs discussion for both time (/shift) invariant (TIV) as well as time (/shift) varying (TV) channels. First of all, in the next section formula to describe the capacity of a TIV AWGN channel will be taken as a starting point and then the capacity of fading channels to be discussed. It is to be noted that only Discrete-Time Systems (DTS) are discussed in this unit. It is assumed that by using sampling most Continuous-Time Systems (CTS) can be converted to DTS. This will help to hold capacity results as derived in DTS. For converting CTS into DTS appropriate sampling rate should be chosen for this conversion.

3.4.1 Capacity Measurements in AWGN Channel

Input-output for the DTS system for AWGN is described as

$$y(n) = x(n) + w(n) \tag{3.5}$$

Where, $x(n)$ and $y(n)$ are real input and output at time n; respectively and $w(n)$ is a zero mean, Gaussian noise that is independent over time.

Considering bandwidth (B), Rx signal power (P), received constant SNR ($\gamma = P/N_0B$), $N_0/2$ being the noise Power Spectral Density (PSD).

Channel capacity by Shannon's formula is

$$C = B\,Log_2(1 + \gamma)\ bits/sec \tag{3.6}$$

• It is interesting to note that Equation (3.6) provides insights into the bandwidth and power. When Bandwidth tends to be very large ($B \to \infty$) Equation 3.6 can be approximated

$$C = B\, Log_2(1 + P/N_0B) \cong Log_2\,(e)\,(P/\,N0) \tag{3.7}$$

In this region, capacity is independent of Bandwidth (B). A simple reason to understand this fact is that available limited power is insufficient to distribute amongst huge available bandwidth.

• When available power is not sufficient power to utilize the available bandwidth., using the approximation that $log(1 + x) \approx x$ for x small yields that, for P small,

$$C = B\, log2\,(1 + p/NoB) \cong log2\,(e)\,(P/\,N0) \tag{3.8}$$

• When P grows large, scaling the bandwidth up by a factor of k leads to the same increase in capacity.

• Capacity limits for AWGN channel (C_{awgn}) with power constraint P and noise variance σ is given as :

$$C_{awgn} = 0.5 * log\,(1 + P/\sigma^2) \tag{3.8}$$

$$C_{awgn} = 0.5 * log\,(1 + P/\sigma^2) \tag{3.9}$$

• Spectral efficiency limits for AWGN channel (C_{awgn}) as f (SNR) is given as:

$$C_{awgn} = log\,(1 + SNR)\, Bits/s/Hz \tag{3.10}$$

When Shannon's theory of information was introduced, the prevailing data rate was of the order of 100 bps. Shannon's capacity theorem was well ahead of that generation and became effective after developments in hardware, modulation, and coding techniques around the 1990s.

Wireless channels typically exhibit either flat or frequency-selective fading depending on available information about the channel. Unlike the AWGN, the capacity of a flat fading channel exhibits different formulas due to dependence on the time-varying channel at the Tx and/or Rx.

3.4.2 Capacity of Flat-Fading Channels

The capacity of this channel depends on the information available about gain $g(i)$ at the transmitter (Tx) and receiver (Rx). There are 3 cases as follows:

- **Fading distributions are known at transmitter and receiver**: Computation of capacity estimates is in this case. Computations are known for only a few models like Finite State Markov channels and Rayleigh fading channels whereby modeling is possible with certain assumptions.
- **Fade value known at only receiver** (via receiver estimation): In this case, Shannon Capacity is given as:

$$C = \int_0^\infty B \, log_2(1+\gamma)p(\gamma)\, d\gamma \;\; bps \tag{3.11}$$

Where $p(\gamma)$ is denoted as the distribution of SNR γ and averaged out over, γ. Jensen's inequality [3] suggests that this capacity is always less w.r.t. AWGN channel for the same mean value of SNR. This points out that there is a decrease in Shannon capacity when fade values are known only at the receiver.

- **Capacity when Fading Known at Tx and Rx** (via receiver estimation and transmitter feedback)

1. If Tx power is fixed, capacity is the same as case number (2).
2. Here, Tx power, as well as Tx rates, are adaptable.
3. The optimal power allocation can be adjustable like (a) Tx power plus Tx rates are increased in case of good channel conditions and (b) Tx power plus Tx rates are decreased in case when it is realized that channel conditions are not good.

3.4.3 Capacity With Receiver Diversity

Receiver diversity is one of the most successfully executed techniques for performance improvement of mobile communications in the presence of fading. The performance improvement due to receiver diversity is achieved as it mitigates signal fluctuations due to fading effects and the channel reshapes again like an AWGN. As receiver diversity overcomes fading effects, curiosity rises to know its effects on the capacity of a fading channel. The quest to find an answer for capacity improvements starts with obtaining the received SNR $p(\gamma)$ distribution for the given diversity combining technique. Deriving the distribution knowledge can be applied to capacity formulas. The specific capacity formula used depends on the assumptions about CSI as described in a couple of paragraphs via 3 different cases. Better the information the more capacity. It is worth noting that as more diversity is created via increasing antenna branches, shaping channels like AVGN, the difference between the capacity values with different formulas also decreases.

3.4.4 Capacity of Frequency-Selective Fading Channels

The capacity of frequency-selective fading channels can be computed as follows:

Time-invariant (TIV) Channels: For TIV channels, capacity achieved remains the same as a flat fading channel but with the distinction that graph changes from Capacity vs. Time to Capacity vs. Frequency.

Time-Varying (TV) Channels: In TV frequency-selective fading channels, impulse response now depends on both frequency as well as time. In this case, effects of self-interference (ISI) will also come into the picture making it difficult to ascertain channel capacity even if H(f, i) is known perfectly. For the cases of Tx and Rx, side information search for an optimal adaptation scheme must consider (a) the past sequence of transmitted bits and their effects and (b) the role of ISI that may affect future transmissions. The capacity estimations of a TV frequency-selective fading channel are difficult to compute in general. However, certain assumptions like approximating this channel by considerations of a group of independent parallel flat fading channels whose resultant capacity is an aggregation of each channel capacity under optimal power allocations.

This unit primarily discussed the capacity computation with Tx and/or Rx having a single antenna. The capacity of MIMO systems with multiple antennas Tx and/or Rx is treated differently and capacity measures for CDMA and MIMO-based systems will be discussed in the next unit.

3.5 MIMO

3.5.1 Introduction

The quality and performance of wireless connections rely on three key parameters: speed (spectrum), range (coverage), and reliability (security). A new radio communication method utilizing multi-dimensional signals has emerged to improve these aspects through a multi-antenna system. The MIMO systems address the challenge of achieving higher data rates, wider coverage, and enhanced reliability without needing an additional spectrum of frequency. When integrated with a multicarrier system, the multi-antenna system delivers exceptional performance.

MIMO plays a crucial role in modern scenarios, particularly in 4G networks. Multiple antenna systems gained popularity about a decade ago for their ability to boost the spectral efficiency of mobile radio systems by utilizing the resource 'space'. A couple of popular examples of open-loop MIMO techniques are Maximum Ratio Combining (MRC), and Space-Time Block Coding (STBC).

Multiple antenna systems serve two main purposes: enhancing link reliability by improving the instantaneous SNR and reducing SNR variations through methods like diversity. MIMO leverages multipath propagation to increase evaluation parameters like (a) throughput, (b) range or coverage, and (c) reliability by transmitting and receiving multiple data signals simultaneously for the specified channel. In a basic smart antenna system, multiple transmitting antennas can transmit the same information, and receiving antennas can receive multiple signals of the same information without changing the transmission capacity. The strongest signal is selected for final reception, known as spatial diversity. Most diversity techniques follow the concept of smart antennas.

The MIMO-based system features two or more Tx antennas and it is different from smart antenna systems. A spatial diversity technique employs multiple Rx antennas. MIMO simultaneously transmits and receives two or more radio signals through a single channel, with each signal carrying distinct information. This enables the system to deliver a data rate per channel that is two or more times greater, known as spatial multiplexing. SDMA-based systems are generally more suitable for this.

MIMO leverages multipath through spatial diversity and the spatial development of multi-antenna systems. The enhanced performance in wireless communication, achieved without requiring additional spectrum (only added hardware and complexity), has contributed significantly to the success of MIMO as a subject for new research. This has led to it being considered as an independent area of study.

MIMO significantly boosts speed to support high-bandwidth applications like streaming multimedia.

3.5.2 Space diversity: concept and system

Within space diversity, signals travel through various propagation paths. In wireless mobile transmission, this can be accomplished through transmit/receive diversity with multiple antennas. Before additional signal processing, a suitable combining mechanism is applied. To study space diversity effectively, it is crucial to categorize four types of wireless systems. Space diversity is further classified into two categories. When antennas are far apart, such as in different BS sites or WLAN access points, it is termed as macro-diversity.

On the other hand, when antennas are within roughly one wavelength's distance, it is known as micro-diversity. Generally, channel reliability is assessed based on Bit Error Rate (BER) performance. In general, four different types of diversity systems can be categorized as follows in Table 3.2.

Table 3. 2 Types of Diversity System

Type of Diversity System	**Diversity Available**
Single input Single Output (SISO)	No Diversity
Single input Multi Output (SIMO)	Receive Diversity
Multi input Single Output (MISO)	Transmit Diversity
Multi input Multi Output (MIMO)	Transmit Receive Diversity

The concept of the SISO system is simply focusing on the communication link between one transmitter and one receiver. Fading significantly affects the possibility of error occurrence in SISO. Additionally, other diversity techniques like SIMO, MISO, and MIMO systems are conceptually illustrated in Figure. 3.5.

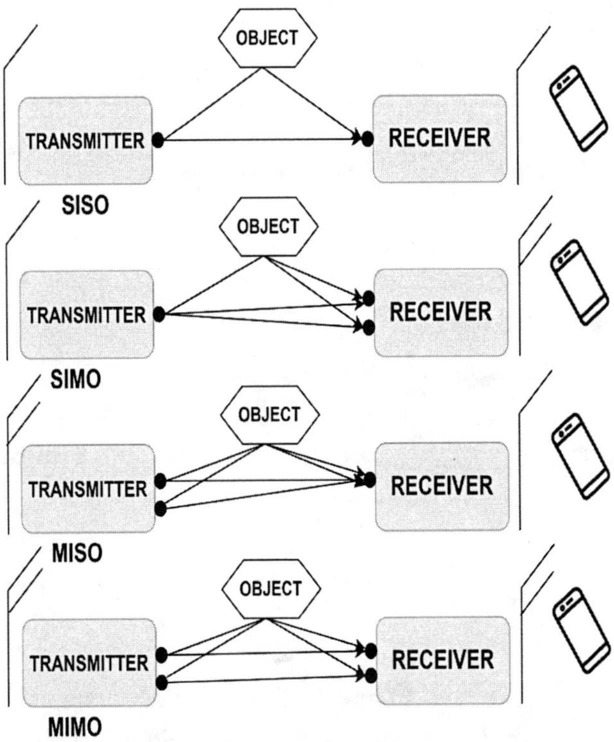

Figure 3.5 Conceptual diagram of different systems along with their spatial diversity

In a SIMO channel, the MRC concept is implemented to leverage diversity at the receiver. The error by using MRC is notably lower compared to the SISO channel. To implement MRC, the receiver must be aware of the fading, meaning access to Channel State Information (CSI) is necessary. Full CSI involves knowing the entire channel transfer function, while partial CSI offers limited channel details, often achieved by transmitting known signals through the channel. Extensive research has been conducted on SIMO radio channel models, focusing on scenarios like cellular environments where mobile transmitters are simple with a single antenna, and base stations have complex receivers with adaptive smart antennas comprising K antenna elements. This scenario is depicted in Figure. 3.6. Scenario in Figure. 3.6 can be conceptualized as the multipath environment in which up to L signals arrive at each BS antenna from different values from (a) no of mobile terminals (l), (b) amplitudes(α_l), (c) phases(ϕ_l), (d) delays(τ_l) and (e) directions (θ_l). These values generally being assumed as Linear Time Invariant (LTI) and, hence Channel Impulse Response (CIR) can be computed by,

$$h(t) = \sum_{l=1}^{L} [\alpha_l(t) e^{j\phi(t)}] \delta[t - \tau_l(t)] a[\theta_l(t)] \tag{3.12}$$

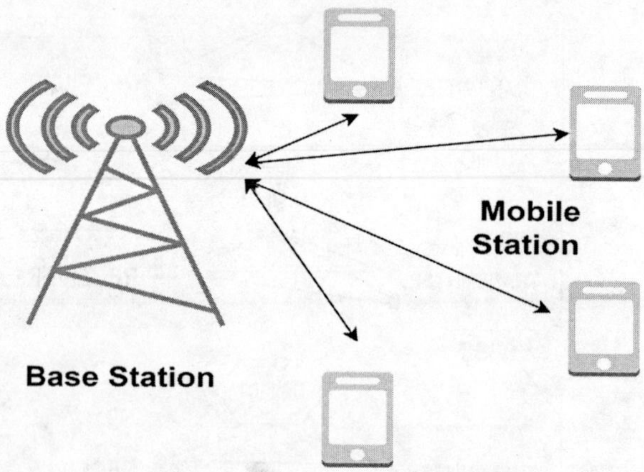

Figure 3.6 SIMO system example

The channel impulse response in mobile wireless links can be represented as a vector sum rather than a simple scalar sum of time. Beamforming, a technique involving the directional reception of waves, can improve the average SNR by concentrating energy in specific directions. However, to execute transmit beamforming successfully; the transmitter needs Channel State Information (CSI) obtained through training sequences, which may decrease throughput. Alternatively, Alamouti's scheme achieves transmit diversity without requiring the transmitter to possess CSI.

Utilizing MIMO channels combines Tx and Rx diversity, offering the benefits of both. This technology can enhance the BER performance of wireless links, tailored to the specific application. MIMO transmits numerous signals simultaneously across the communication channel, boosting communication capacity. Spectral efficiency, a common metric for wireless capacity, can be enhanced by transmitting multiple signals with distinct information streams over the same frequency channel.

In a TDD system, the transmitter has complete CSI, rendering the channel reciprocal. In FDD, the transmitter provides partial CSI, such as system information or the most robust Eigen mode of the channel. Transmit diversity allows for a resilient wireless link even without CSI at the transmitter, achievable through space-time codes like the Alamouti code in Multiple-Input Single-Output (MISO) systems.

Although CSI-based transmitter optimization has benefits, it is not practical to expect the receiver to obtain accurate CSI for the transmitter. Usually, the transmitter has a noisy channel estimate or a quantization index that approximates the actual channel.

3.5.3 Smart Antenna and MIMO

A smart antenna is a wireless system strategically developed to assist operators in managing fluctuating traffic volumes and network inefficiencies effectively. This innovative system enables carriers to adjust gain settings, thereby expanding or contracting coverage in specific localized areas without the need to physically ascend a tower or install an additional customized antenna. This flexibility allows carriers to adjust a cell's coverage to match its unique traffic patterns and adjust its operations according to different parameters like as the time of day, day of the week, or periods of high traffic.

Smart antenna systems can broadcast and receive signals in a way that is both adaptive and directionally (spatially) aware by combining an antenna array with sophisticated digital signal processing capabilities. These systems may dynamically change the directionality of the radiation pattern in response to the surrounding signal environment, which greatly improves wireless system performance. Smart antenna systems work similarly, to how our ears and brain identify the source of sound when a user is moving within a base station's coverage area. Switched-beam systems and adaptive-array systems are the two main forms into which these systems are usually divided. Both strategies seek to actively decrease noise or interference coming from other directions while directing a primary lobe or radio beam in the direction of a specific user. MIMO (Multiple Input, Multiple Output) and smart antenna systems share a commonality in utilizing multiple antennas. However, they differ significantly in their fundamental principles. Smart antennas function by enhancing traditional radio systems through the optimization of signal energy along the primary path, maximizing signal reception strength in real time. They use both beamforming (i.e. Transmitter diversity) as in Figure. 3.7 and receive combining as in Figure. 3.8. To serve as methods to mitigate multipath interference. The combination of both beamforming and receive combining techniques as shown in Figure. 3.9 presents motivation to improve spectral efficiency.

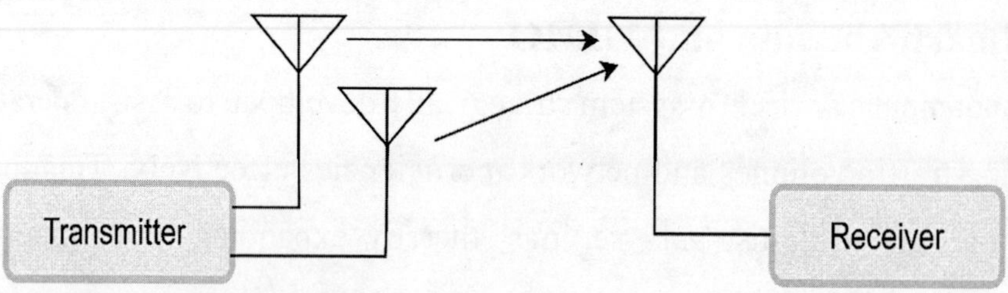

Figure 3.7 Beamforming (beam steering) with two transmitting antennas.

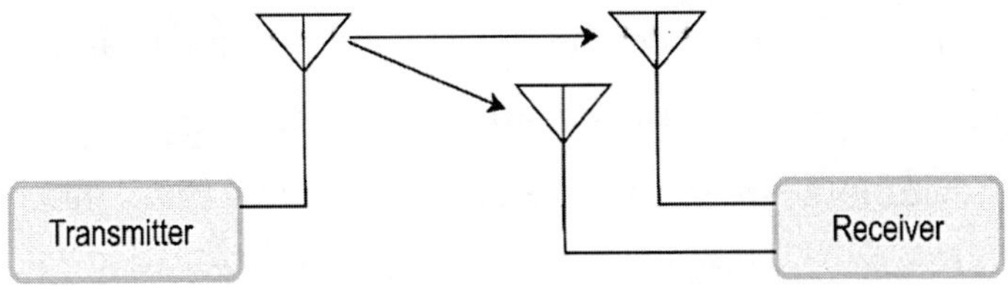

Figure 3.8 Diversity with two receiving antennas

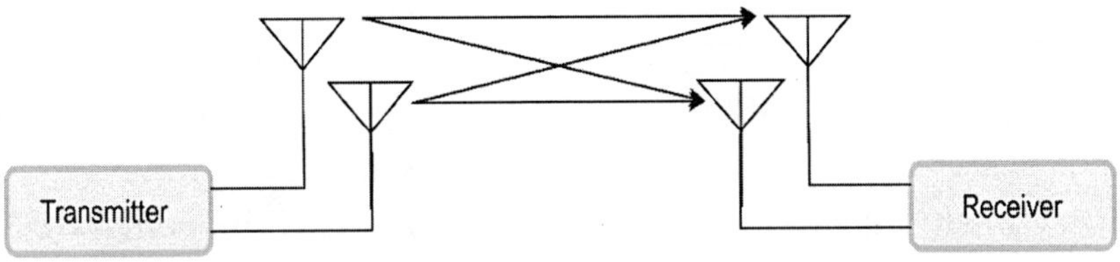

Figure 3.9 Combination of Beamforming (beam steering) and Diversity in smart antenna.

Using multiple antennas is a frequent feature of both smart antenna systems and MIMO (Multiple Input, Multiple Output) systems. However, their underlying ideas are very different from one another. By optimizing signal energy along the principal path and maximizing signal reception intensity in real-time, smart antennas, improve conventional radio systems. To reduce multipath interference, they employ both beamforming (also known as transmitter diversity), as shown in Figure. 3.8, and receive combining, as shown in Figure. 3.9. The utilization of beamforming and receive combining techniques, as demonstrated in Figure 3.10, offers incentives for enhancing spectral efficiency.

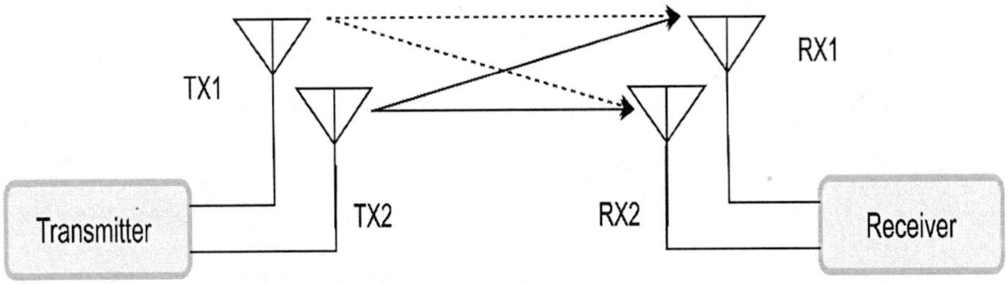

Figure 3.10 MIMO using Multi Tx- Multi -Rx to send a multi-signal on the same channel

MIMO's mathematical foundations entail data transmission across a matrix channel as opposed to a vector channel, which opens up possibilities beyond array gain or diversity advantages alone, like improved spectrum efficiency.

MIMO systems, in contrast, resemble an extension of smart antennas, similar to CDMA transmission, in which several users share a single time/frequency channel. However, by introducing distinct input stream signatures, or "virtual users," in an essentially orthogonal manner without frequency spreading, MIMO gradually improves point-to-point application performance.

Table 3.3: Comparison between the Smart Antenna and MIMO.

Smart Antenna	**MIMO Technology**
1. In the smart antenna technique, either the transmitter/ receiver or both of them are equipped with more than one antenna. (Usually, more cost and space are easily available at the BS end). 2. The data is transmitted over a vector channel. 3. Normally, few smart antennas perform better in LOS or close to LOS systems.	1. In MIMO both the ends have more than one antenna. (The subscriber end is a more sophisticated device with a maximum number of features included rather than just a pocket telephone. 2. The data is transmitted over a matrix channel. 3. MIMO performs well in non-LOS and it tries to mitigate the multipath rather than exploiting it.

Multi-Input several Output, or MIMO, technology allows several streams to be coded and decoded simultaneously for a single goal. For spatial selectivity, a rich multipath is needed. According to research on smart antennas, frequency spectrum utilization and data rate can be increased by installing an array of antennas on the transmitter or reception end. Fourth-generation wireless networks might not, however, be able to meet consumer demands for multimedia and e-commerce services.

3.5.4 MIMO-Based Architecture

The concept of MIMO systems can be elucidated as follows. In a wireless mobile communication system, envision a scenario where both the transmitting and receiving ends are furnished with multiple antenna elements. This configuration is depicted in Figure. 3.11, featuring N transmitting antennas and M receiving antennas. A fundamental tenet of MIMO systems involves space-time signal processing, a technique that augments the temporal dimension inherent in digital communication data with the spatial dimension facilitated by multiple spatially distributed antennas. This approach enables the conversion of a single incoming message sequence into parallel sequences that can be independently processed. The standard required blocks include (a) source coding, (b) channel coding, (c) modulation, and (d) RF up-conversion at the sending end, and their counterparts at the receiver end, potentially tailored for individual antenna elements or employing adequate signal-processing techniques. Subsequently, the operations of space-time encoding and decoding are elaborated.

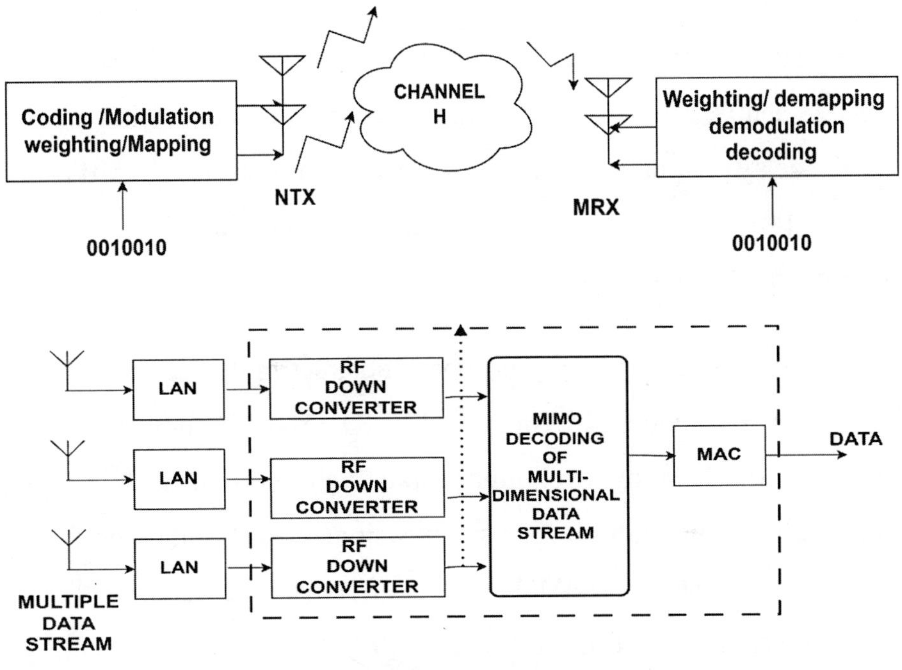

Figure 3.11 A simple MIMO wireless transmission system
a) Conceptual view b) Important blocks in a MIMO RF system.

At the receiving end of a MIMO system, signal processing is employed to disentangle the multiple streams. For instance, in a 3 x 3 MIMO system, often regarded as a 'three-measurement three-unknown problem,' the successful resolution of this challenge relies on each receiving antenna capturing a distinct combination of transmitted symbols. Failure to achieve this condition results in an unsolvable system of linear equations due to the channel matrix lacking full rank, with antennas exhibiting strong correlations influenced by factors such as spacing, polarization, and radiation patterns.

3.5.5 MIMO- a Solution for Multipath

Multipath signals denote transmitted signals that traverse diverse routes to reach the receiver, including reflections off objects, the ground, or atmospheric layers. Despite their potential, these signals can sometimes be weakened or disregarded, resulting in energy wastage. Notably, strong multipath signals, if excessively potent, can compromise the efficiency of wireless equipment. Graphically, radio signals can be visualized as sine waves denoting amplitude over time. When a delayed multipath signal coincides with the primary signal, its peaks and troughs may misalign, inducing signal attenuation and distortion. In cases where the delay multipath signal (secondary signal) are peaks aligned with the primary signal's troughs (areas between two peaks), partial or complete cancellation of the main signal may occur as shown in Figure. 3.12.

Conventional radio systems typically do not counteract multipath interference or utilize specific mitigation techniques. In contrast, MIMO leverages multipath propagation to enhance data transmission by exploiting parallel channels. This technology involves transmitting and receiving multiple data streams simultaneously through distinct antennas within the same radio channel. MIMO systems effectively combat signal fading, thereby enhancing reliability and performance.

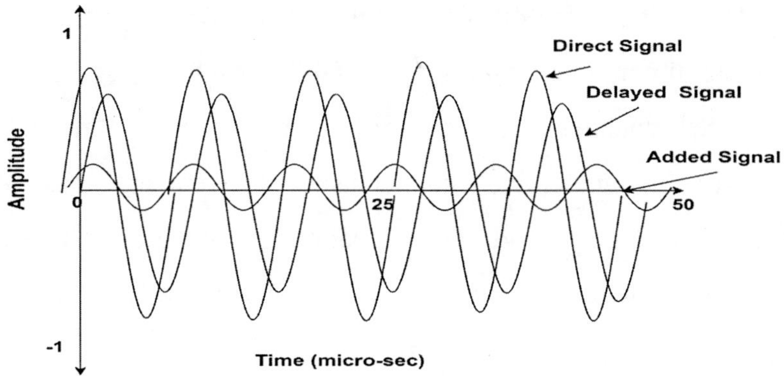

Figure 3.12 Subtractive addition of multipath signal resulting in direct, delayed (slightly attenuated) and added (heavily attenuated) signal

Radio signals can be illustrated on a graph as a sine wave (y-axis amplitude and the X-axis time) as in Figures 3.12 and 3.13. It is evident that when a multipath signal arrives delayed w.r.t. Primary one in such a way that their peaks and troughs do not align perfectly, resulting in a weakened and blurred combined signal received by the receiver. When the multipath signal's peaks coincide with the primary signal's troughs due to sufficient delay, it can partially or completely cancel out the main signal. Traditional radio systems may either not address multipath interference or rely on the primary signal to overpower the interfering signals, or they may use techniques to mitigate multipath issues by adjusting the delays of received signals to realign the peaks and troughs.

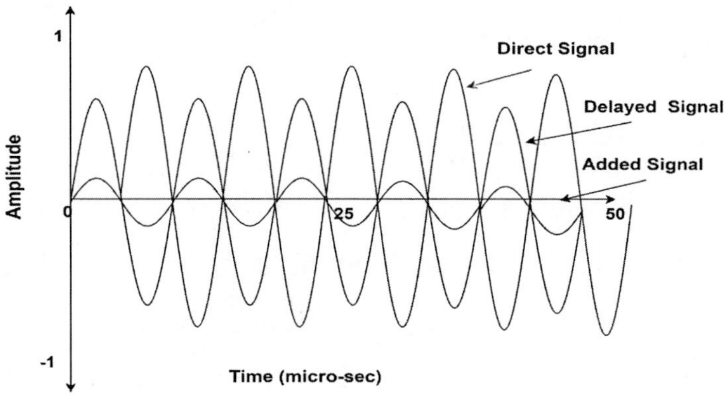

Figure 3.13 Subtractive addition of multipath signal resulting in direct and reflected signals with a large difference

This concept is also utilized in MIMO systems, where MIMO capitalizes on multipath propagation. Instead of combating multipath signals, MIMO harnesses them to carry more data through parallel channels. By transmitting and receiving multiple data signals simultaneously over the same radio channel using different antennas, MIMO exploits multipath signals to enhance information capacity. As mentioned earlier, fading is a significant challenge in wireless channels, and MIMO addresses this issue effectively, resulting in significantly improved reliability.

3.5.6 Multipath Diversity Reception in Multi-Antenna Receiver

The strength of the signal received relies on the carrier frequency, as the wireless channel acts as the dielectric medium where waves travel at varying phase velocities, reaching the receiver at different times. A signal sent at a specific carrier frequency and moment might arrive at the receiver in a multipath null. Utilizing diversity reception lessens the likelihood of communication failures caused by fades by consolidating multiple copies of the same message from various channels. The effectiveness of diversity techniques diminishes if signal fading is correlated across different branches, emphasizing that the received signals should ideally be uncorrelated. Illustrated in Figure 3.14 is a one M-branch receiver displaying maximum ratio combining, where phase shifters and attenuators work to align signals in terms of phase and amplitude.

3.5.7 MIMO Channel Capacity

Earlier sections detailed wireless systems that utilized space diversity techniques having multiple receive antennas to encounter multipath fading issues. Fading was considered a factor that diminishes system quality, requiring the implementation of space diversity at the receiving end to counteract it. Here, we delve into Multiple-Input Multiple-Output (MIMO) wireless communications, also known as Multiple-Tx, Multiple-x (MTMR) in academic literature.

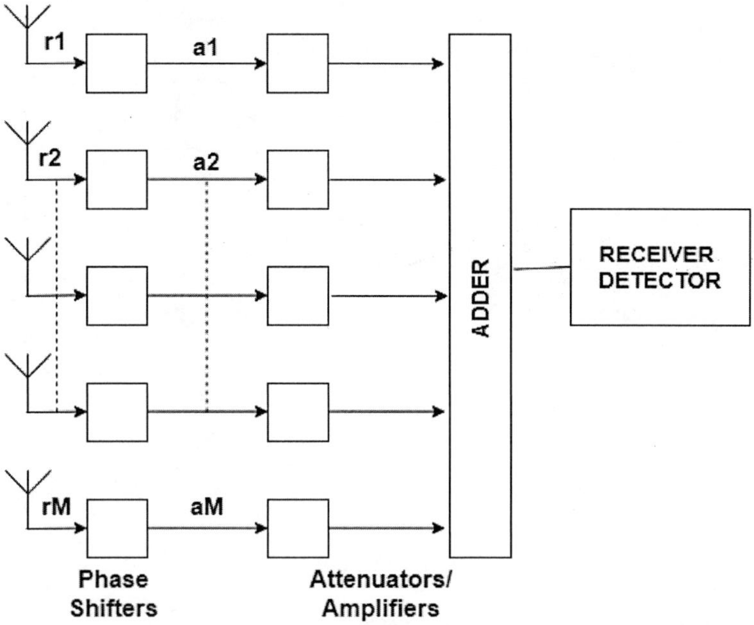

Figure 3.14 A design of M-M-branch antenna diversity receivers

The MIMO-based system employs space diversity at the receiver with the inclusion of the following three points:

- The fading effect is considered as a source of possible enrichment factor rather than as a negative influencer.
- Space diversity utilized at TX and Rx ends provides the possibility of improvement in (a) channel capacity or (b) spectral efficiency.
- Unlike conventional techniques, MIMO aims to achieve capacity increase by increasing computational complexity but retaining parameters like total $P_{transmit}$ and $BW_{channel}$ fixed.

Figure 3.15 illustrates MIMO wireless connections in block diagram form. The signals sent by N transmit antennas across the wireless channel are selected to be within the same frequency band. As expected, the channel scatters the transmitted signals differently. Additionally, because of transmissions of multiple signals, the system encounters a spatial type of signal-dependent interference known as Co-Antenna Interference (CAI).

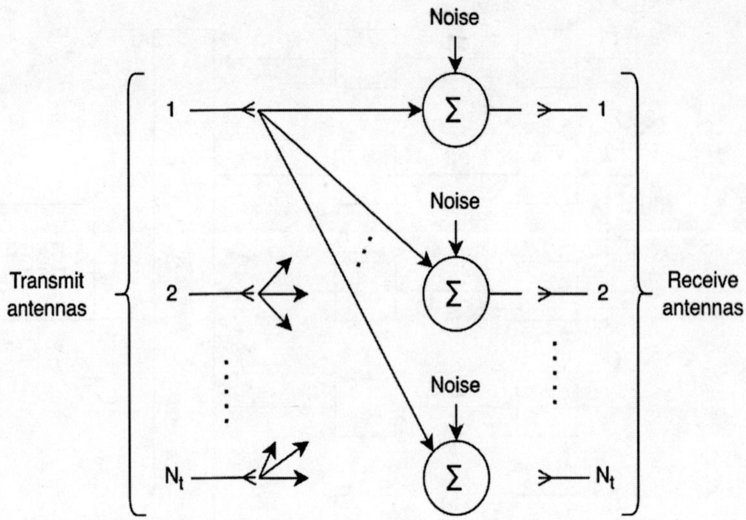

Figure 3.15 Block diagram of MIMO wireless link.

3.5.8 Basic Channel Model

In a MIMO system having narrowband signal transmission and operating over a flat-fading channel, assume that there are N_t transmit antennas and N_r receive antennas as shown in Figure 3.16. This antenna setup is denoted as a pair(N_t, N_r). To conduct a statistical analysis of the MIMO system, a baseband representation of the Tx and Rx signals, along with the channel is used. Specifically, the following notation is introduced for clarity:

The spatial parameter (N) is defined as,

$$N = min\{N_t, N_r\} \tag{3.13}$$

Depicts parameters for a MIMO channel with N_t Tx-antenna and N_r Rx- antennas.

The N_t-by-1 vector

$$s(n) = \left[\underline{s_1}(n), \underline{s_2}(n), \ldots\ldots \underline{s_{Nt}}(n)\right]^T \tag{3.13}$$

Denotes the signal $s(n)$ with N_t dimensional vector (equal to several antennas) at time instant$n$. The symbols $s(n)$ are assumed to have a mean equal to zero and common varianceσ_s^2. The total power of transmission (P) is fixed and defined as

$$P = N_t\sigma_s^2 \tag{3.14}$$

To keep P constant, σ_s^2 must be inversely proportional to N_t.

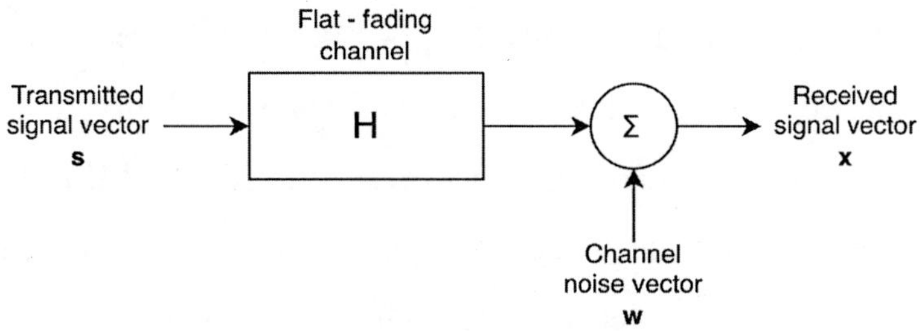

Figure 3.16 Basic channel model

The propagation environment and successful MIMO performance hinge on the correct implementation of the antenna system heavily influences the MIMO channel's capacity. It is crucial to note that both the propagation environment and the array configurations affect the transfer matrix H. The challenge lies in determining the most effective array topology for maximizing capacity or minimizing symbol error rates, as this choice is contingent on the specific characteristics of the site's propagation. One general guideline is to position antennas as far apart as feasible to diminish the correlation between received signals. As the array size increases and antennas are placed closer together, the simulated capacity continues to rise, while the measured capacity per antenna decreases due to increased correlation between neighboring elements.

3.5.9 Outage Capacity

Another commonly used measure of channel capacity is outage capacity. Capacity is viewed as a random variable that relies on the channel's instantaneous response and remains steady throughout the transmission of a coded block of information of finite length. Therefore, it represents the MIMO channel's short-term behavior achieved through coding within the fading interval. If the channel's capacity drops below the outage capacity threshold, there is no chance that the transmitted block of information can be decoded without errors, regardless of the coding scheme employed.

3.5.10 Advantages of MIMO

MIMO technology provides various advantages, such as reduced power usage and cost, extended range and coverage, enhanced link quality and reliability, and improved spectral efficiency. By directing transmissions towards specific users, MIMO can consume less power and reduce expenses towards higher amplifications. The array or beamforming gain amplifies signal strength at the receiver by combining signals received from all antenna elements coherently. Moreover, MIMO can enhance data rates and spectral efficiency by utilizing spatial multiplexing gain, by having concurrent transmission of multiple messages through spatial signatures.

3.5.11 MIMO Applications

Several techniques are being viewed as complementary to MIMO to enhance throughput performance and spectrum efficiency, generating interest, especially in improving 3G mobile systems like HSDPA. While MIMO offers increased throughput in ideal urban channel conditions with uncorrelated fading between antennas, it is sensitive to varying channel conditions. MIMO elements are recognized as suitable for MIMO, and MIMO-OFDM technology significantly enhances wireless LAN performance, supporting various applications more efficiently and enabling new demanding applications. Traditionally, increasing speed involves sacrificing range and reliability, extending range comes at the expense of speed and reliability, and improving reliability requires reducing speed and range. However, MIMO-OFDM technology revolutionizes these trade-offs by enhancing all of these parameters at the same time. Although MIMO will benefit all major wireless industries, the wireless LAN sector is at the forefront of leveraging MIMO advancements. Consequently, MIMO-OFDM serves as the foundation for IEEE 802.11n standard proposals. MIMO techniques provide substantial enrichment to the usage of various wireless systems including cellular networks, Wi-Fi, WiMAX, mobile satellite TV, satellite radio and RFID.

Wi-Fi

Small devices that primarily operate indoors typically prefer adaptive arrays due to several advantages:

- Enhanced range
- Improved interference mitigation, especially in unlicensed band settings
- Consistent coverage
- Leveraging MIMO technology becomes an appealing solution to achieve higher data rates.

WiMAX

To increase capacity and range, base stations employ adaptive arrays and multi-beam antennas. WiMAX users employ adaptive arrays to make up for lost building penetration. Equitable gains in both directions are guaranteed by time-division duplexing.

Cellular

Concepts of multi-beam and adaptive arrays are helpful at base stations (BS) in cellular networks, whilst adaptive arrays are useful for smartphones. Higher data rates, more system capacity, and wider coverage are the main benefits that follow.

RFID

When applied to readers, smart antennas like multi-beam or adaptive arrays can increase the range at which an RFID response can be detected.

UNIT SUMMARY

A thorough examination of the vital role diversity plays in contemporary wireless communication networks is given in this session. It starts by explaining why diversity is so important for reducing the negative impacts of fading, interference, and other channel impairments and improving signal performance and dependability.

Subsequently, the module explores several kinds of diversity, such as polarization, frequency, time, and spatial diversity.

Using thorough explanations, people get knowledge regarding the various tactics utilized to counteract signal deterioration, thereby guaranteeing strong communication connections.

Space diversity approaches are a major focus of this module since they are essential for mitigating the effects of spatial fading. There is a detailed discussion of methods including Selection Combining, MRC, EGC, SLC, and MIMO antenna systems. Every strategy is discussed to clarify its guiding principles, benefits, and drawbacks, allowing students to comprehend how it can be used in real-world situations.

Additionally, the subject discusses the important topic of antenna systems' capability for space diversity. Students are given the information they need to optimize system capacity and spectrum efficiency by exploring concepts like radiation patterns, antenna gain, and spatial multiplexing.

To sum up, this subject equips students with the fundamental knowledge and abilities needed to plan, evaluate, and maximize diversity-related ideas for the construction of reliable wireless communication networks.

Students who comprehend the value of diversity, its several forms, and the nuances of space diversity strategies will be more equipped to meet the demands of contemporary communication networks.

Exercises

Multiple Choice Questions

1. Why are adaptive equalizers necessary in mobile communications?
   a. To reduce fade depth and duration in flat fading scenarios.
   b. To compensate for the anticipated range of channel amplitude and delay properties.
   c. To address inter-symbol interference (ISI) in multipath settings.
   d. To employ diversity through multiple receiving antennas.

2. Which diversity technique leverages the decorrelation of signals with horizontal and vertical polarization to achieve diversity gain?

   a. Polarization diversity
   b. Time diversity
   c. Frequency diversity
   d. Angle diversity

3. What is a characteristic of frequency diversity?

   a. It involves transmitting signals at varying time intervals.
   b. It relies on the spacing of antennas to achieve diversity gain.
   c. It utilizes multiple independent fading carrier frequencies.
   d. It spaces out transmissions beyond the channel's coherence time.

4. What is the primary purpose of spatial diversity in wireless communication?

   a. To reduce interference from adjacent channels.
   b. To mitigate fading effects by utilizing multiple antennas.
   c. To increase the modulation performance of the system.
   d. To minimize the number of handoffs between cells.

5. Which diversity combining method coherently aligns all signals with equal gain for all signal levels to achieve the highest SNR at the receiver?

   a. Maximal-ratio combining
   b. Selection diversity
   c. Square-law combining
   d. Equal Gain Combining (EGC)

Answers to Multiple Choice Questions

1) c 2) a 3) c 4) b 5) d

Short Questions

1. Explain the role played by equalization in mobile communications. Explain the necessity of adaptive equalizers.

2. Define diversity. Enlist various types of diversities.

3. Enlist various advantages offered by diversity.

4. Define the following type of diversity:
   - o Antenna diversity
   - o Frequency diversity
   - o Time diversity
   - o Polarization diversity
   - o Code Diversity

5. Explain in brief any three types of diversity.

6. Explain in brief: Coherence Bandwidth and Coherence Time

7. Briefly explain selection diversity.

8. Describe the principle behind Equal Gain Combining (EGC) in wireless communication systems and explain its advantages compared to other diversity combining methods.

9. What is square-law combining?

Long Questions

1. Explain the concept of spatial diversity and its significance in wireless communication systems.

2. Explain the concept of maximal ratio Combined with the help of a suitable diagram.

3. Explain the concept of MIMO antenna systems capacity with the help of a suitable diagram.

4. Explain the concept of spatial diversity in wireless communication systems and provide examples of scenarios where space diversity is particularly effective.

PRACTICAL

1. Simulate MIMO system in Downlink using N Input and M Output Antennas. Differentiate between the performance of MIMO and Massive-MIMO based on the performance parameters.

KNOW MORE

Advancements in MIMO architecture for 6G provide significant benefits over current systems. The field of study on intelligent reflecting surfaces (IRS) is appealing. These surfaces, which also include tiny antennas, have the potential to affect radio waves by reflecting them in specific directions. This allows for improved signal quality for individual users, increased coverage in areas with low signal penetration, and more concentrated signal transmission. Moreover, the potential of a family of synthetic materials called met surfaces to dynamically control electromagnetic waves is being studied. Researchers hope to create smart antennas that can adjust their emission patterns in real-time by integrating met surfaces into MIMO arrays. As a result, extremely effective beamforming would be possible, maximizing signal strength for particular users and reducing interference for others. Future MIMO systems will be complicated; hence, artificial intelligence (AI) must be incorporated. To optimize beamforming techniques and resource allocation and guarantee the most effective utilization of the network's capacity, AI algorithms can assess user demands and network conditions in real time. With the use of AI, creative antenna design, and higher frequency bands like terahertz (THz), 6G will be able to provide users with previously unheard-of levels of data speeds, network flexibility, and dependability. To optimize beamforming techniques and resource allocation and guarantee the most effective utilization of the network's capacity, AI algorithms can assess user demands and network conditions in real time. With the use of AI, creative antenna design, and higher frequency bands like terahertz (THz), 6G will be able to provide users with previously unheard-of levels of data speeds, network flexibility, and dependability.

Dynamic QR Code for further Reading

REFERENCES AND SUGGESTED READINGS

[1.] Theodore Rappaport - Wireless Communications, Principles and Practice-ISBN 0130422320 (Edition) PHI)

[2.] NPTEL course: https://archive.nptel.ac.in/courses/117/102/117102062/

[3] Goldsmith A. Capacity of Wireless Channels. In: Wireless Communications. Cambridge University Press; 2005:99-125.

4 Overview of CDMA, OFDM and LTE

UNIT SPECIFICS

In this unit, the following aspects are discussed:

- *Spread spectrum, its characteristics and types*
- *CDMA, its types, capacity, channels, power control, call processing etc.*
- *OFDM and its block diagram, synchronization, efficiency*
- *Overview of LTE*

RATIONALE

In the last couple of decades with the increased demands of mobile wireless systems, issues of limited spectrum efficiency and multipath propagation have also increased. To tackle such issues, this module starts with Spread *spectrum techniques, and their characteristics, types including direct sequence spread spectrum,* and frequency hopped Spread Spectrum. Intending to increase wireless network capacity and provide a good alternative to existing time and frequency-based multiple access techniques, the unit also builds basic knowledge about CDMA, its *types, channels, call processing etc.* It covers a discussion about the capacity of CDMA channels and a comparison with GSM. The module also presents the concept of orthogonality and examines the suitability of OFDM *as a* multi-carrier system. It also covers *block diagrams of OFDM, its synchronization and efficiency.* At last, this module discusses the evolution of LTE systems that will be helpful as an essential tool for designing and maintaining current-generation mobile wireless systems.

PRE-REQUISITES

Mathematics (Class XII)
Basics of Signals and Wave
Basics of Communication system

UNIT OUTCOMES

After going through this unit, students will be able to:

***U4-O1**: Explain the concept of spread spectrum, including its characteristics, types -DSSS and FHSS.*

***U4-O2**: Explain CDMA, capacity, advantages of CDMA channels, power control, call processing and CDMA standards: IS-95, CDMA2000 and WCDMA.*

***U4-O3**: Explain the OFDM with the help of a suitable block diagram...*

***U4-O4**: Describe long-term evolution and all IP networks*

Unit-4 Outcomes	***MAPPING of UNIT OUTCOMES WITH COURSE OUTCOMES*** *(1- Weak Correlation; 2- Medium correlation; 3- Strong Correlation)*				
	CO-1	***CO-2***	***CO-3***	***CO-4***	***CO-5***
U4-O1	2	--	1	2	3
U4-O2	1	1	--	1	3
U4-O3	1	--	--	3	3
U4-O4	1	--	--	2	3

4.1 Introduction to CDMA signal generation

This module delves into the concepts of CDMA signal generation, Walsh codes, PN codes, channel capacity and CDMA transmitter and receiver systems. CDMA systems transcend multiple access mechanisms, incorporating fundamental technologies within their framework. At the core of CDMA, functionality lies in the employment of codes to facilitate the conversion of analogue voice signals into digital formats, segregating voice and control data into distinct channels. By employing spreading codes, often derived from PN sequences, basic voice signals are transformed into digital signals, creating discrete channels. Before we start CDMA, the concept of spread spectrum modulation is presented to address issues of limited spectrum efficiency and multipath propagation issues.

4.1.1. Spread Spectrum Modulation

In Spread Spectrum (SS) modulation, the message signal is combined with other sequences in such a way that the resultant sequence requires a larger bandwidth i.e. spectrum is spreaded. The use of SS helps designers overcome problems with limited spectrum efficiency and multipath propagation in wireless system designs. Use SS techniques were originally designed for military communications but are now widely employed in commercial applications, particularly mobile radio networks and satellite communication. SS transmissions have three major advantages over fixed frequency transmissions namely (a) they offer higher SNR (b) due to longer sequence their interception by unauthorized persons is difficult, and (c) minimal interference with conventional frequency bands. In addition, they also achieve goals like lesser requirement of energy spectral density, better multiple-access facility, better security, anti-jamming capabilities, multipath protection, and routing.

Spread spectrum systems can be divided into averaging and avoidance systems. Averaging systems minimize interference over a long period, whereas avoidance systems avoid interference during a short period. Although SS is a wideband modulation technique with a high bandwidth expansion factor, it does not resist white noise in the same way that FM, or PCM.

The following two requirements apply to the spread spectrum system:

- The bandwidth used for transmission is far greater than the bandwidth used for sending the data.
- A special (pseudo-noise-like) code is mixed at the transmitter to achieve spectrum spreading and for perfect retrieval; the same code is to be used at the receiver.

The following are some of the traits of the spread spectrum:

- A greater capacity for the channel
- Resistance against the spread of multiple paths
- Difficulty of interception by unauthorized parties
- Robustness in the case of the possibility of a jamming problem
- Immunity to multipath propagation distortion
- Multiple access capabilities are supported

4.1.2. Types of Spread Spectrum Technique

There are two primary methods for spreading the spectrum:

1. Spread Spectrum with Frequency Hopping (FHSS)
2. Spectrum of Direct Sequence Spreading (DSSS)

4.1.2.1. Frequency hopping spread spectrum (FHSS)

The frequency or channel rapidly changes in a FHSS, as illustrated in Figure. 4.1, while the transmitter switches between channels in an established pseudo-random sequence. The pseudo-random sequence generator and the list of channels (the hop set) in the receiver are the same as those in the transmitter.

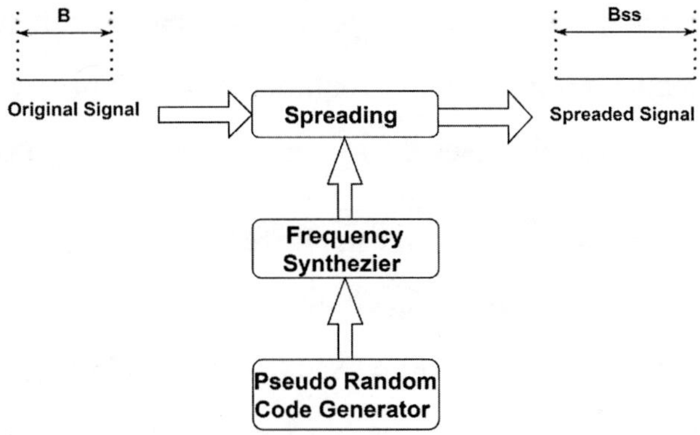

Figure 4.1 Conceptual diagram of Frequency Hopping Spread Spectrum (FHSS)

(B-is BW of the source sequence and Bss-bandwidth of the spreaded sequence)

A synchronizing circuit in the receiver that offers anti-jamming capabilities guarantees synchronization between the pseudo-random code generators in the transmitter and receiver.

The frequency or channel rapidly changes in a FHSS, as illustrated in Figure. 4.1, while the transmitter switches between channels in an established pseudo-random sequence.

The pseudo-random sequence generator and the list of channels (the hop set) in the receiver are the same as those in the transmitter.

A synchronizing circuit in the receiver that offers anti-jamming capabilities guarantees synchronization between the pseudo-random code generators in the transmitter and receiver.

4.1.2.2. Advantages and disadvantages of FHSS

When compared to the Direct Sequence Spread Spectrum (DSSS), FHSS has a larger processing gain and is less dependent on synchronization with distance. FHSS does, however, have certain drawbacks.

One potential drawback of the system is its high bandwidth requirements, which are frequently in the gigahertz region. It also requires the use of pricey and sophisticated digital frequency synthesizers, which raises the system's total cost and complexity.

4.1.2.3. Direct sequence spread spectrum (DSSS)

The Direct Sequence (DS) technique is employed to modulate the data. In DSSS, the user signal is multiplied by a high-bandwidth pseudo-noise code sequence.

This code sequence, also known as the chip sequence due to its chip rate occurrence, contributes to the modulation process.

The modulated signal is then transmitted across the radio channel. DS plays a key role in expanding the bandwidth. The below Figure. 4.2 shows the DSSS conceptual block diagram.

Using the logic illustrated in Figure.4.3, the unique Pseudo Noise (PN) code, which to an unauthentic user seems to be a random sequence, is employed in the DSSS to spread and de-spread the signal.

That is, the output for logic 1 is an inverted PN sequence, and the output for logic 0 is the same sequence because of an XOR operation.

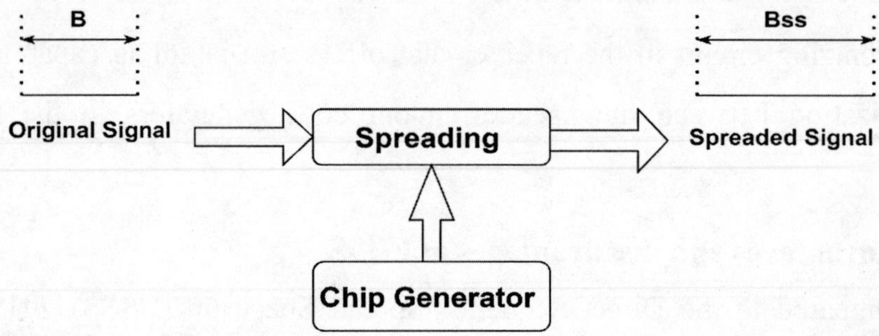

Figure 4.2 Conceptual diagram of Direct sequence Spread Spectrum (DSSS).

In the direct sequence system, a clock rate is used to generate each bit of a PN code, enabling phase shift keying of the data and expanding its bandwidth. This system employs a double-balanced mixer driven by the PN code to alternate the phase of a carrier between 0 and 180 degrees, known as BPSK.

Unlike a frequency-hopping transmitter, the direct sequence signal is created directly by a long-duration PN code that repeats the same cycle. The DS receiver spreads the wideband signal by utilizing a synchronized pseudo-random sequence identical to the one in the transmitter.

To achieve synchronicity, the receiver needs a clock rate adjustment circuit to align its PN code with that of the transmitter and match the hopping pattern.

A tracking mechanism is essentially required to sustain synchronism, once established. Other spread spectrum modulation techniques include Time-Hopping Spread Spectrum (THSS), hybrid methods, and Chirped Spread Spectrum (CSS).

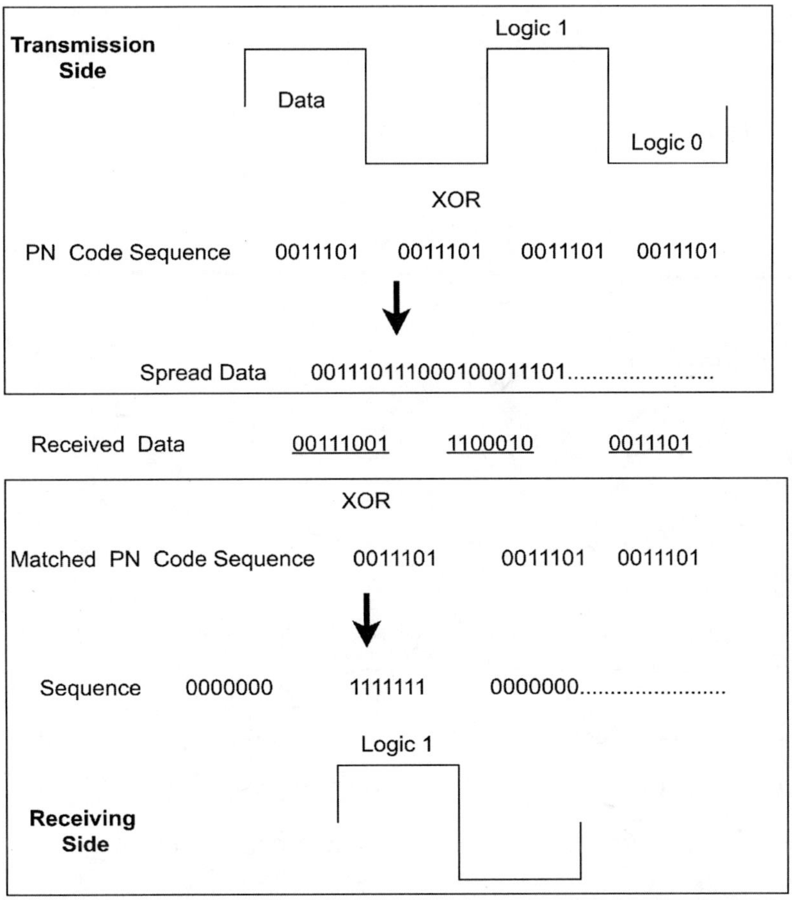

Figure 4.3 Illustration of transmission of '10' code (1 followed by 0) using DSSS

4.1.2.4. Advantages and disadvantages of DSSS

When compared to the FHSS, Direct Sequence Spread Spectrum performs better in the presence of noise and tackles jamming issues in a better way. DSSS reduces signal interference from other sources. However, there are disadvantages to DSSS as well. Compared to FHSS, it has a smaller channel bandwidth and a lesser processing gain. Furthermore, different transmitter and receiver distances might interfere with synchronization, making regular communication difficult.

4.1.2.5. Relation between spread spectrum bandwidth and signal to noise ratio

Let us rewrite Shannon's system capacity (C) as

$$C = W \, log_2 \left(1 + \frac{S}{N}\right) \tag{4.1}$$

$$\frac{C}{W} = log_2 \left(1 + \frac{S}{N}\right) \tag{4.2}$$

Or

Changing bases using $log_a b = \frac{log_e b}{log_e a}$ and $log_2 e = \frac{1}{log_e 2}$

$$\frac{C}{W} = log_2 e \; log_e \left(1 + \frac{S}{N}\right) \tag{4.3}$$

$$\frac{C}{W} = 1.44 \; log_e \left(1 + \frac{S}{N}\right) \tag{4.4}$$

$$\cong 1.44 \; log_e \left(\frac{S}{N}\right)$$

By logarithmic expansion,

$$log_e \left(\frac{S}{N}\right) = \frac{S}{N} - \frac{1}{2}\left(\frac{S}{N}\right)^2 + \frac{1}{3}\left(\frac{S}{N}\right)^3 - \frac{1}{4}\left(\frac{S}{N}\right)^4 \ldots\ldots \tag{4.5}$$

In a spread spectrum system, the signal-to-noise ratio is typically small and much less than 0.1. Hence, higher-order terms than first order are neglected.

Therefore, taking an approximation for the wider bandwidths

$$\frac{C}{W} \cong \frac{1.44\, S}{N} \quad \text{Thus, } W \cong \frac{C}{1.44\,(S/N)} \tag{4.6}$$

It is evident from the obtained relationship that raising the transmission bandwidth will result in the lowering required signal-to-noise ratio for a fixed data rate C.

4.1.3. Types of Codes used in CDMA Signal Generation

In general, CDMA belongs to two primary groups: synchronous (orthogonal codes) and asynchronous (pseudorandom (PN) codes). Let us look at each one separately.

4.1.3.1. Walsh code

For CDMA applications, Walsh Codes are most frequently utilized as the orthogonal codes. These characters represent the lines of a unique square matrix known as the Hadamard matrix. It consists of n lines to produce a square matrix of $n \times n$ Walsh codes for a set of codes of length N.

Walsh functions a 64 in the IS-95 system. This matrix contains all zeros on the first line, and distinct combinations of bits 0 and 1 are present on each of the subsequent lines. Every line represents binary bits equally and is orthogonal to other lines. Each mobile user employs one of the available row sequences in the matrix as a spreading code while using the CDMA system. Furthermore, it offers no cross-correlation with any other user. The following is the recursive definition of this matrix:

$$W_1 = [0]$$

$$W_{2n} = \begin{bmatrix} W_n & W_n \\ W_n & \overline{W_n} \end{bmatrix}$$

Where n assumes values in power of two decides dimensionality of the matrix W. Moreover, n denotes the NOT logic operation on every bit in this matrix. The Walsh function for dimensions 2, 4, and 8 is displayed by the three matrices W_2, W_4 and W_8, in that order.

$$W_2 = \begin{bmatrix} 0 & 0 \\ 0 & 1 \end{bmatrix}$$

$$W_4 = \begin{bmatrix} 0 & 0 & 0 & 0 \\ 0 & 1 & 0 & 1 \\ 0 & 0 & 1 & 1 \\ 0 & 1 & 1 & 0 \end{bmatrix}$$

$$W_8 = \begin{bmatrix} 0 & 0 & 0 & 0 & 0 & 0 & 0 & 0 \\ 0 & 1 & 0 & 1 & 0 & 1 & 0 & 1 \\ 0 & 0 & 1 & 1 & 0 & 0 & 1 & 1 \\ 0 & 1 & 1 & 0 & 0 & 1 & 1 & 0 \\ 0 & 0 & 0 & 0 & 0 & 0 & 0 & 0 \\ 0 & 1 & 0 & 1 & 0 & 1 & 0 & 1 \\ 0 & 0 & 1 & 1 & 0 & 0 & 1 & 1 \\ 0 & 1 & 1 & 0 & 0 & 1 & 1 & 0 \end{bmatrix}$$

4.1.3.2. Pseudo noise (PN) codes or direct sequence (DS) fundamentals

In this scheme, the basic message signal (baseband signal) is multiplied by the PN sequence of high bandwidth. The result of multiplication is the PN code that is transmitted over the communication channel. PN sequence is utilized to generate hopping frequencies or hopping time slots.

4.1.3.3. Properties of PN codes

The randomness of a PN sequence is assessed by examining certain code properties observed throughout a complete period. We shall discuss these properties one by one.

- **Balance property:** where the number of one in each sequence period exceeds the number of 0s by exactly one digit.

Take, for example, a standard PN code: 0001 0011 0101 111

Number of ones = 8

Number of Zeros= 7

It can be said that a given code meets the balance property to qualify for the PN code.

- **Run Length Property**: It is preferred that roughly half of the runs in each period consist of ones and zeros, with one-half of each type being of length 1, one-fourth of length

2, one-eighth of length 3, and so forth. Let us revisit the same code 0001 0011 0101 111 for further analysis.

Number of runs =8

000	1	00	11	0	1	0	1111
3	1	2	2	1	1	1	4

It can be observed that out of a total of eight runs; four runs are of length 1, two runs are of length 2, and one run each is for length 3 and 4. It can be concluded that a given code meets the run length property to qualify for PN code.

- **Autocorrelation Property:** For valid PN sequences, its autocorrelation function is periodic. When it is executed this function value can be defined as:

$$No.of\ disagreements\ (d) = No.of\ Disagreements\ (a) + 1$$

For comparison mentioned below, the First row represents the original sequence while the second row represents one digit-shifted sequence

```
0 0 0 1 0 0 1  1 0 1 0 1 1 1
1 0 0 0 1 0 0  1 1 0 1 0 1 1
-----------------------------
d a a d d a d a d d d  d a a a
```

$$No.of\ disagreements\ (d) = 8$$

$$No.of\ Disagreements\ (a) = 7$$

It can be concluded that a given code meets the autocorrelation property to qualify for PN code.

4.1.3.4. Maximum length (ML) sequences

Maximum Length Shift Registers (MLSR) is a Linear Feedback Shift Register (LFSR) that generates an ML sequence that is used as desired PN codes subject to meeting certain criteria. Generated PN codes are periodic in nature and for p stages in LFSR, the length of the ML sequence will be 2^{P-1}. If this length is kept large and also meets PN properties as mentioned earlier it will appear random to users. Figure. 4.4 shows Galois implementation consisting of shift registers and modulo two sums.

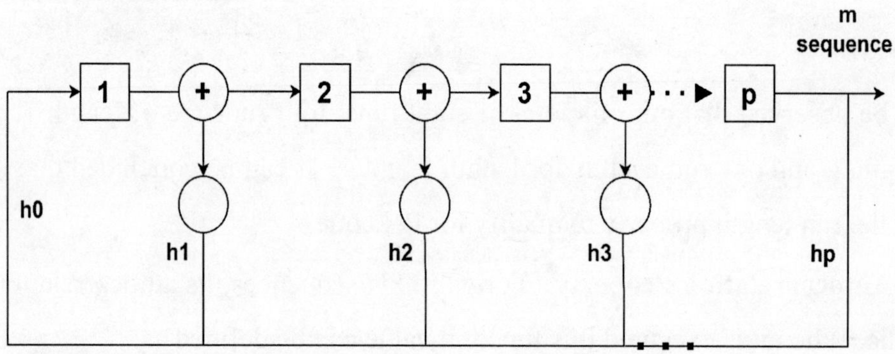

Figure 4.4 LFSR using Galois implementation and modulo two sum

In general, the non-reducible primitive polynomial $h(x)$ as described in Equation 4.7 is implemented using LFSR.

$$h(x) = h_p x^p + h_{(p-1)} x^{(p-1)} + \ldots + h_1 x^1 + 1 \tag{4.7}$$

$h(x)$ Is called a generator polynomial with coefficients h_i either take the value 1 (feedback connection present) or zero (feedback connection not present) with two extreme coefficients h_0 and h_p always being one (connection present) so that polynomial order remains maximum (p). The initial state of the shift register also should be a primitive polynomial $q(x)$ that generates the PN sequence along with generator polynomials $h(x)$. It should be noted that $q(x)$ and $h(x)$ should be irreducible polynomials to LFSR resulting in MLSR that will in turn generate a maximum length sequence used as PN code.

4.1.4. Capacity of CDMA System

CDMA is an interference-limited system. In a given transmission between a base station and a given user, all other signals are considered as noise. This fact points out that as long as noise coming from surrounding users is acceptable we can increase the number of users i.e. capacity of the system. To find the approximate capacity (C) of the CDMA system, let us define interference noise spectral density (I_0) as:

$I_0 = Noise\ power\ spectral\ density$

$$I_0 = \frac{Noise\ Power}{Available\ bandwidth} \tag{4.8}$$

Now, let us assume that the power (P) received by each user is the same and there are M users in the system. We note that except the desired single user, all other users (i.e.$(M-1)$ users) are generating noise, each with power received as P,

$$Noise\ power = (M-1)P \tag{4.9}$$

Considering available bandwidth(W), Equation 4.8 can be rewritten as,

$$I_0 = \frac{(M-1)P}{W} \tag{4.10}$$

Rearranging terms above Equation 4.10 can be written as,

$$M-1 = \frac{I_0\, W}{P} \tag{4.11}$$

Now, power P is defined as energy per bit (E_b)multiplied by bit rate (R) i.e.

$$P = E_b.R \tag{4.12}$$

Substituting (4.12) in the Equation (4.11), we get

$$M-1 = \frac{I_0\, W}{E_b.R} \tag{4.13}$$

Or

$$M-1 = \frac{\frac{W}{R}}{\frac{E_b}{I_0}} \tag{4.14}$$

$M >> 1$ We can write $M-1 \cong M$

Therefore, we can write

$$M = \frac{\frac{W}{R}}{\frac{E_b}{I_0}} \tag{4.15}$$

4.1.5. Comparison of CDMA and GSM Capacity Tradeoffs

Equation 4.15 is the basic approximated capacity equation for CDMA systems. It is important to note that parameter $\frac{E_b}{I_0}$ can be interpreted as BER and the following interpretations can be derived about the comparison of capacity of CDMA and GSM.

(1) If BER ($\frac{E_b}{I_0}$) is constant and W is constant; capacity is inversely proportional to Bit rate R. The lower the bit rate, the higher the capacity of the CDMA system. Such a tradeoff is not available in GSM; lowering bitrate can increase the quality of transmission in GSM, not its capacity.

(2) If $\frac{W}{R}$ is constant, a reduction in BER increases the capacity of the CDMA system. Again, increasing BER can increase the quality of transmission in GSM, not its capacity.

(3) If BER and Bit rate are constant then M is directly proportional to W i.e. bandwidth of the CDMA system.

4.1.6. General Block Diagram of DS Transmitter and Receiver

On the transmitter side, MOD 2 adders are utilized for biphasic and quadriphase modulation, along with balanced modulators and carriers. This is conceptually shown in Figure 4.5. For biphasic modulation, the transmitter employs an MOD 2 adder and for quadriphase modulation, two MOD 2 adders are used, supplied with carriers that are 90 degrees phase-shifted.

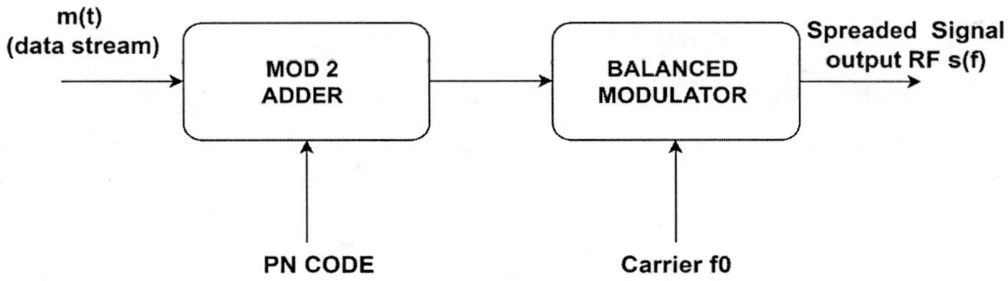

Figure 4.5 Carrier modulation and spreading of original data to get SSM signal.

The SSM RF output is accessible, and binary data and code are employed for message modulation. The receiver identifies signals, eliminates carriers, and despread using PN sequences through active and passive techniques.

The receiver layout is depicted in Figure. 4.6. A spread signal receiver must execute three key functions: detecting the signal presence, removing the carrier, and despreading or demodulating utilizing PN sequences. The detection and despreading process can be active or passive. Active methods seek the signal's presence in both time and frequency domains and follow the sequence post-acquisition, while passive methods only necessitate searching for the signal in terms of frequency. The decision between active and passive methods is influenced by sequence length and processing gain. Active methods monitor sequences, while passive methods react to signals in terms of frequency. Integrating both strategies in a receiver can improve performance, as seen in active rake receivers with antenna diversity.

The selection of methods hinges on the circumstances, with active methods favoured for lengthy sequences and significant processing gain, whereas passive methods might be preferable for shorter sequences or as an acquisition aid. Rake receiver designs based on active methods with antenna diversity are presently being advanced to enhance performance.

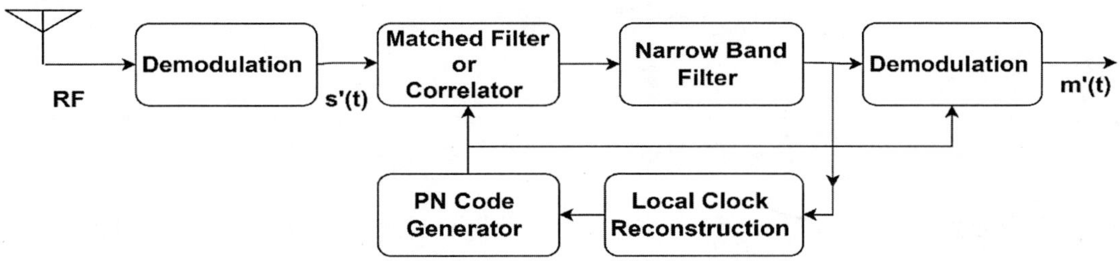

Figure 4.6 Carrier demodulation and despreading of SSM signal to get the original data.

4.1.7. Near-Far Problem and Power Control

The near-far issue in mobile systems refers to the challenge of strong signals overwhelming weaker signals at the receiver. This problem is especially pronounced in DSSS systems. To illustrate this problem, consider a scenario with a receiver and two transmitters (see Figure. 4.7). One transmitter (A) is near the receiver, while the other (B) is far away. When both transmitters simultaneously emit signals with equal power, the receiver picks up more power from the closer transmitter (A), creating difficulty for the farther transmitter (B). As one transmission acts as noise for the other, the farther transmitter requires a significantly higher signal-to-noise ratio. If the closer transmitter's signal strength greatly surpasses that of the farther transmitter, the SNR for the latter may fall below the necessary threshold, rendering its signal indistinguishable and essentially halting its transmission. This results in a communication channel jam. To ensure successful communication, the far transmitter would need to substantially boost its transmission power, which may not be feasible. In essence, the near-far problem involves identifying and capturing a weaker signal amidst stronger signals. Additionally, rapid signal level fluctuations due to fading must be addressed to maintain consistent received signal strength at the base station, necessitating the use of power control techniques in CDMA systems.

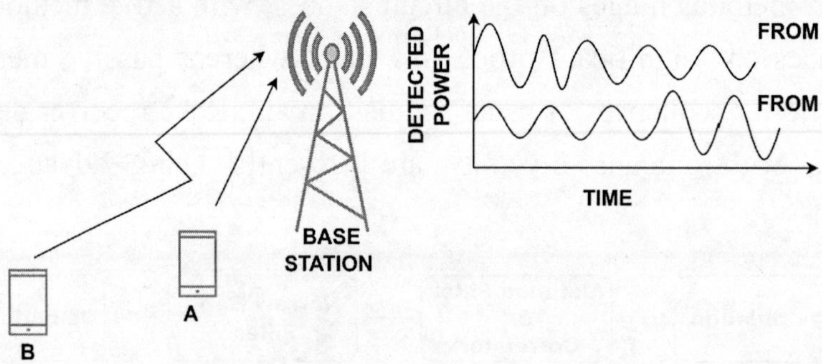

Figure 4.7 Near Far Problem Scenario

In CDMA systems or similar cellular networks, this challenge is commonly addressed by dynamically adjusting the output power of transmitters. Closer transmitters operate at lower power levels to maintain a balanced SNR for all transmitters at the receiver. As all mobile devices transmit on the same carrier frequency, internal interference within the system significantly affects system capacity and voice quality. Two key conditions must be met simultaneously:

- Each mobile device must regulate it's transmit power to control interference.
- The power level should ensure satisfactory voice quality.

The primary goal of power control is to manage transmitted power on both forward and reverse links while upholding link quality across all conditions (refer to Fig. 4.8).

In closed loop system power control, as shown in Fig. 4.8

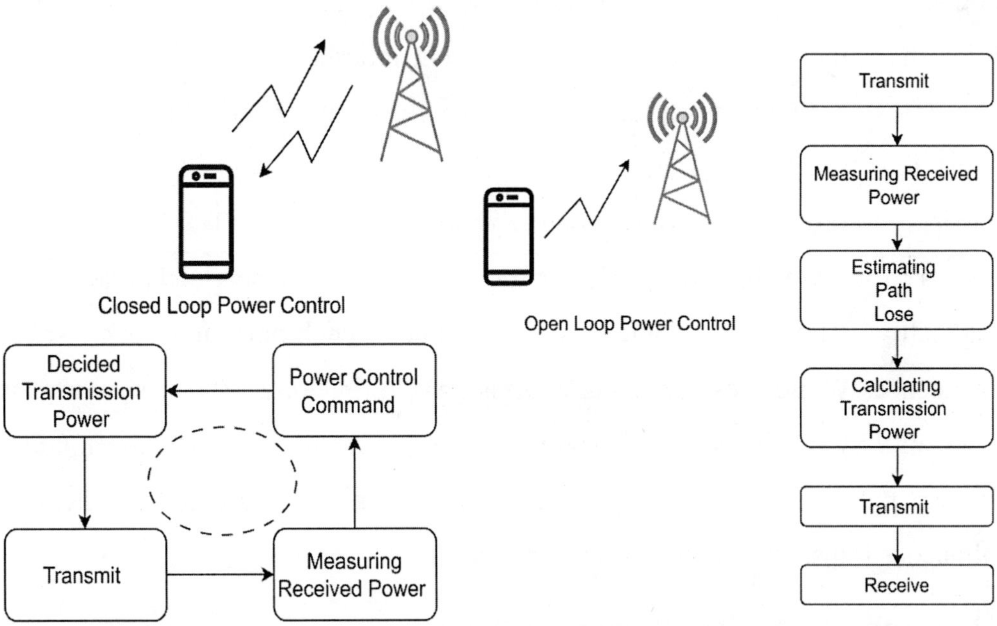

Figure 4.8 Illustration of Closed and Open Loop Power Control System in Mobile

First transmission power related commands are generated by BS and passed to MS. Subsequently, based on that command received, MS decides the power level and sends further signals to BS. BS computer power signal level received from given BS and generates power control command for next iteration and process repeats.

In open loop control systems, Signals transmitted from one end are received from the other end, path losses are estimated and based on that transmitted power levels are calculated. Estimated Power control effectively counteracts fading fluctuations by equalizing received power from all mobile stations. By implementing power control, the near-far problem is alleviated, resulting in comparable detected power levels for users A and B as depicted in Figure. 4.7. In contrast, a frequency hop system discussed later is less susceptible to the near-far problem as it operates as an avoidance system rather than an averaging system.

4.2 Overview of OFDM

4.2.1. Basic Principle of Orthogonality

Orthogonality refers to the independence of two signals during a specific timeframe, enabling their transmission and detection without interference on a shared resource i.e. channel. This concept is essential for understanding the Orthogonal Frequency Division Multiplexing (OFDM) system, which achieves orthogonality by closely spacing subcarriers to conserve spectrum. Time Division Multiplexing (TDM) allocates distinct time slots to individual data signals for transmission over a single channel to maintain orthogonality. However, challenges like time synchronization can affect this independence. While most Frequency Division Multiplexing (FDM) systems maintain orthogonality by spacing signals in frequency, they consume more spectrum. OFDM sets itself apart by assigning each symbol or group of symbols to a unique carrier, enabling simultaneous transmission through multiple carriers.

4.2.2. Single Vs. Multi-Carrier Systems

Multicarrier systems like OFDM were developed in the 1950s and 1960s but faced implementation delays due to technological challenges. The introduction of the Fast Fourier Transform (FFT) calculation algorithm in 1965 by Cooley and Tukey enabled effective implementation, allowing multiple carriers to mitigate channel effects through diversity. Single-carrier frequency allocation in conventional communication links is depicted in Figure 4.9.

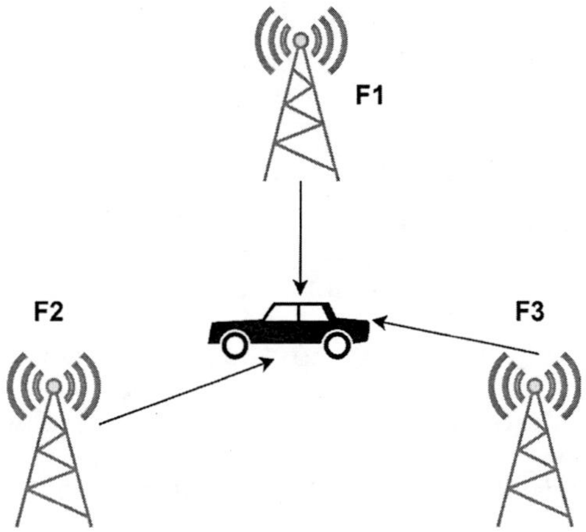

Figure 4.9 Single Carrier Allocation of Frequency to the Users in a network.

4.2.2.1. Single-carrier systems

Efficient utilization of the frequency spectrum in single-carrier systems is very challenging. It requires the placement of modulated message signals very close without Inter-Carrier Interference (ICI) problems by using Frequency Division Multiplexing (FDM). In the ideal case, each message bandwidth should be exactly adjacent to the neighbor's message with spectrum utilization requirement, but in actual cases, certain guard bands are essential to attenuate adjacent message signals. Usage of guard bands results in wastage of bandwidth. Transmitting information at high data rates requires shorter symbol periods. The symbol period is inversely proportional to the baseband data rate ($T = 1/R$). Therefore, as R increases, the symbol period T decreases, and vice versa. In a multipath environment, shorter symbol periods can lead to increased ISI. A shorter symbol may in turn increase spectral efficiency and data rate but may increase the chance of symbol overlap. M-ray schemes with higher values of M provide better spectral efficiency and data rate at the expense of SNR.

4.2.2.2. Multicarrier systems

OFDM, a multicarrier system, addresses the challenges faced by single-carrier systems discussed above. OFDM segregates the available spectral bandwidth into more subchannels (or subcarriers). By keeping all subchannels narrowband, they experience nearly flat fading, simplifying equalization. OFDM allows overlapping of resultant message bandwidths without interference and exerts bandwidth plus spectral efficiency improvements relative to single-carrier counterparts. To enhance spectral efficiency, subchannel frequency responses overlap and remain orthogonal. This orthogonality is maintained, even in time-dispersive channels, by introducing a cyclic prefix. OFDM supports high data rates by converting symbols from serial to parallel, resulting in longer symbol durations that help mitigate ISI. However, time and frequency synchronization remain the primary limitations of multicarrier systems.

4.2.3. OFDM Signal Mathematical Representation

The arithmetic below is for a lattice-type OFDM structure, where it is simpler to insert a pilot for channel estimation. A hybrid multiple-access multi-carrier modulation method is the OFDM. N_c subcarriers separated by the frequency distance Δf make up an OFDM signal. Thus, N_c equidistant subchannels make up the entire system bandwidth W. Every subcarrier has a time interval of length $T_S = \frac{1}{\Delta f}$ and is mutually orthogonal. This reduces the impact of both ISI and ICI concurrently by making efficient use of the spectrum and transmitting N_c streams with orthogonal carriers.

The Kth subcarrier signal is described analytically by the function $g_k\,(t)$

$$g_k\,(t) = \{e^{j2\pi\Delta ft} \quad 0 < t < T_s\ , k = 0, \dots\dots N_c - 1\ 0 \quad otherwise \tag{4.16}$$

In comparison to a single-carrier transmission system spanning the same bandwidth W, the OFDM symbol duration T_S is N_c times longer since the OFDM system bandwidth W is divided into N_c narrowband subchannels. The number of subcarriers is selected for a particular system bandwidth so that the symbol duration is greater than the channel's maximum delay. The signal $g_k\ (t)$ is a subcarrier signal that is prolonged by a cyclic prefix (guard interval) with length T_g to prevent ISI.

$$g_k\ (t) = \{e^{j2\pi\Delta ft} \quad 0 < t < T_s + T_g\ , k = 0, \ldots .N_c - 1\ 0 \quad otherwise \qquad (4.17)$$

Only the time interval $[0, T_s]$, also known as symbol time, is evaluated at the receiver once the guard interval is eliminated. The guard interval is therefore a complete system overload. is $T = T_s + T_g = base\ period$ is the overall OFDM lock duration. There is no ISI and the subcarriers' orthogonality is unaffected if the guard interval length T_g is greater than the radio channel's maximum delay.

While the multipath channel modifies the sub-carriers phase and amplitude during the assessed time interval, interference with the prior information only manifests itself during the guard interval. A certain range of phases can be used to guide decision-making when it comes to phase recovery. The following provides a more accurate frequency and temporal domain representation of the OFDM signal.

Single real OFDM subcarrier

$$S(k) = e^{j\theta_m}\ \delta\left\{k - m - \frac{N}{2}\right\} + e^{-j\theta_m}\ \delta\left\{k + m - \frac{N}{2}\right\} \qquad (4.18)$$

Composite (real) OFDM subcarrier

$$S(k)_{ofdm} = \sum_{m=c\ first}^{c\ last} \quad e^{j\theta_m}\ \delta\left\{k - m - \frac{N}{2}\right\} + e^{-j\theta_m}\ \delta\left\{k + m - \frac{N}{2}\right\} \qquad (4.19)$$

4.2.4. Selection Parameters for Modulation

These four prerequisites are necessary for the OFDM system to function:

- **Available bandwidth:** When choosing several subcarriers, the bandwidth limit will be a major factor. Many subcarriers with appropriate CP lengths can be obtained with a lot of bandwidth.

- **Required bit rate**: The system must be able to deliver the data rate needed for the particular application.
- **Maximum allowable delay spread** particular to a user environment should be specified in advance when calculating the CP length.
- **Doppler values:** It is important to consider the impact of the Doppler shift brought on by user movement.

Let W be the total bandwidth that is accessible. Assume that the channel's maximum delay spread is T_d seconds. Select a guard interval T_s for the OFDM symbol that is significantly larger than the maximum delay spread T_d, such as $T_g = 4 \times T_d$, to avoid ISI. Select an OFDM symbol time T_s that is significantly larger than the guard time T_g, for example, $T_s = 8 \times T_g$, where T_s is symbol time without guard interval, to lessen the cost caused by the cyclic prefix. Subcarrier spacing $\Delta f = \frac{1}{T}$ for $T = T_s + T_g$. $N_c = \frac{W}{\Delta f}$, or the nearest power of two, is the number of subcarriers

The symbol duration should be significantly longer than the guard period to reduce the signal-to-noise ratio (SNR) loss caused by the guard time. Long-duration symbols, however, are vulnerable to phase noise, frequency offset, and Doppler dispersion.

The computations above lead to the following two conclusions:

1. The frequency gap between subcarriers decreases with increasing symbol duration. Thus, more subcarriers can be supported for a given signal bandwidth. On the other hand, extending the symbol duration reduces the signal bandwidth for a given number of subcarriers.
2. There are more samples per OFDM symbol when there are more subcarriers. It does not, however, necessarily follow that the symbol duration lengthens. The time interval between two samples gets shorter if the OFDM symbol duration stays constant. This suggests that the bandwidth of the OFDM signal has increased. However, if the bandwidth of the OFDM

signal is fixed, then adding subcarriers will result in a decrease in the frequency spacing between two subcarriers, which will lengthen the symbol.

4.2.5. Windowing in OFDM and Spectral Efficiency

The side lobes and main lobes will appear in the OFDM spectrum as a whole due to the sinc form of each narrowband channel. These should be removed from the final transmission bandwidth since they are undesired elements. Applying these windowing functions to the final OFDM baseband signal, which contains the effects of every subcarrier, will accomplish this. Certain mathematical functions define, for instance, the Keiser, Hanning, Bartlet, and Hamming windows.

OFDM baseband signal transmission Fig. 4.10 Out-of-band components can be eliminated by using the windowing block to apply the windowing function to the OFDM. For this reason, it can also be thought of as a shaping filter. Dark lines in Figure. 4.10 show the concept of windowing. This will enable retaining the orthogonality and allowing the next transmission bandwidth to be designed more in line with the prior one.

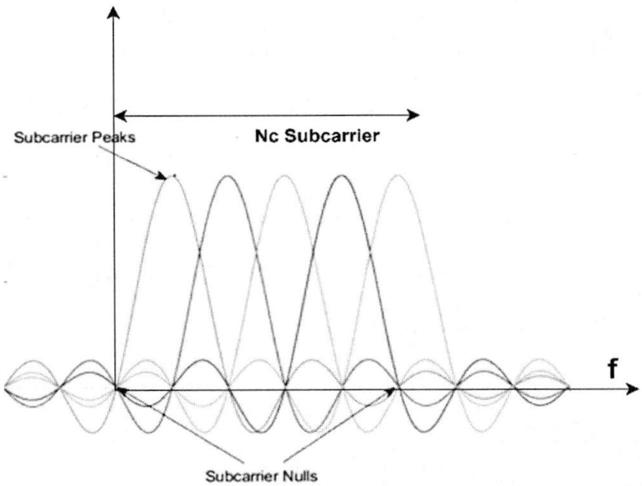

Figure 4.10 Application of Windowing Function to the OFDM baseband Signal.

4.3 Overview of Long-Term Evolution (LTE)

4.3.1. Introduction

In comparison to 3G technologies, LTE is a 4G wireless standard that offers mobile computing devices faster and more network capacity. Carriers employ this technology to send wireless data to customers' phones, enabling wireless broadband communication for mobile devices. LTE offers greater speed, efficiency, peak data rates, and frequency and bandwidth flexibility compared to the prior 3G version.

With peak data transfer rates of up to 100 Mbps downstream and 30 Mbps upstream, LTE outperforms 3G in this regard. It offers lower latency, expandable bandwidth, and backward compatibility with both the Universal Mobile Telecommunications Service (UMTS) and Global System for Mobile Communication (GSM) technologies. LTE-Advanced (LTE-A) is a later developed version of LTE and it can produce a peak speed of about 300 Mbps.

LTE is technically slower than 4G but still faster than standard 3G, even though it is frequently referred to as 4G LTE. True 4G offers speeds up to 1,000 Mbps, whereas LTE only reaches 100 Mbps. Different LTE variants, such as LTE-A, can reach 4G speeds, nevertheless. Eventually, LTE spread over the world as a standard and widely used in places without 5G. The development of the 5G standard is influenced by LTE. To handle 5G data sessions, early 5G networks—known as non-standalone 5G - need a 4G LTE control plane. 4G network architecture may be used to construct and sustain 5G networks to save operators' capital and operating costs as they roll out 5G.

4.3.2. Working of LTE

The Third Generation Partnership Project (3GPP) created LTE. The standard was referred to as the next development in mobile telecommunications, building upon the requirements of the GSM and UMTS networks. Originally, LTE was not considered to be true 4G. 4G was first described by the International Telecommunication Union (ITU) as a cellular standard that would provide 100 Mbps for mobile users and 1 Gbps for stationary users. Later, 4G to LTE was applied in addition to several other wireless standards.

For its downlink signal, an LTE network uses OFDMA, a multiuser variation of OFDM. With better spectral efficiency than 3G, OFDMA allows the LTE downlink to send data at faster data rates to numerous users from a BS. The uplink signal is transmitted using single-carrier FDMA, which lowers the MS's required transmit power. Transmission Control Protocol/Internet Protocol (TCP/IP) upon which the upper layers of LTE are built, produces an all-IP network akin to wired communications. Data transmissions including mixed data, phone, video, and messaging traffic are supported by LTE. MIMO antenna, common to the IEEE 802.11n wireless local area network standard, is employed in LTE-A. Higher SNRs at the receiver are made possible by MIMO and OFDM, which enhance wireless network coverage and throughput, particularly in crowded urban locations. Devices using LTE-A must have a specific chip installed. Qualcomm, Nvidia, and Broadcom are a few of the manufacturers of LTE-A-compatible processors. Most smartphones are compatible with LTE-A. Different countries saw the deployment of LTE at different times from telephone carriers. North American and some European operators introduced roughly between 2009-2012. The average 4G network availability across many countries reached 80% and a few larger mobile carriers like AT&T, T-Mobile and others registered 4G availability at 90% and higher in/ after 2020. In North America and Western Europe, 3G is being replaced by 4G LTE and 5G, with the closure of major 3G networks expected to occur by 2025. The Global Mobile Suppliers Association (GSA) estimated two-thirds of all mobile customers worldwide were connected by LTE by now. For a better overview, a few facts about LTE can be noted:

- **Successor Technology**: LTE is the technology that comes after CDMA 2000 and UMTS.

- **Performance Improvement:** LTE offers greatly greater spectral efficiency and up to 50 times better performance for cellular networks, which is a considerable improvement.
- **Greater Data Rates:** With peak downlink speeds of 300 Mbps and peak uplink speeds of 75 Mbps, LTE seeks to attain greater data rates. Data rates with a 20 MHz carrier can surpass 300 Mbps in ideal circumstances.
- **Support for High Data Rates:** Voice over IP (VoIP), streaming multimedia, video conferencing, and high-speed cellular modems all require high data rates, which LTE is perfect for meeting.
- **Time Division Duplex (TDD) and Frequency Division Duplex (FDD)** are two duplex modes used by LTE. TDD employs the same frequency spaced apart by time, whereas FDD uses separate frequencies for uplink and downlink communications.
- **Flexible Carrier Bandwidths**: LTE is compatible with both FDD and TDD modes and can support carrier bandwidths ranging from 1.4 MHz to 20 MHz. The frequency range and the network operator's available spectrum determine the precise bandwidth that is used.
- **MIMO Technology:** several Input Multiple Output (MIMO) transmissions are supported by all LTE devices, enabling base stations to broadcast several data streams simultaneously over the same carrier.
- **IP-Based Interfaces:** Every LTE network interface, including the backhaul links to radio base stations, is IP-based. In contrast to previous methods that relied on expensive, narrowband E1/T1 and frame relay links, this is simpler.
- **Quality of Service (QoS)**: Even in situations when network capacity is limited, standardized QoS procedures guarantee that voice call requirements—such as consistent latency and bandwidth—are satisfied.

- **Compatibility and roaming:** LTE uses both new and existing spectrum for 2G and 3G as well as UMTS and GSM, EDGE, and UMTS systems. Both handover and roaming to current mobile networks are supported.

4.3.3. Benefits of LTE

- **High Throughput:** High downlink and uplink data rates are made possible by LTE, which leads to high throughput.
- **Low Latency:** Power-saving states can be rapidly entered and departed, increasing efficiency, and connection durations are slashed to a few hundred milliseconds.
- **FDD and TDD on the Same Platform:** LTE provides flexible deployment options by supporting both FDD and TDD on a single platform.
- **Better End-User Experience**: Mobility management, air interface protocols, and connection establishment signaling optimization improve user experience. The reduction of latency to 10 ms enhances user satisfaction even further.
- **Smooth Connection:** LTE ensures compatibility and continued service by enabling smooth connections to current networks such as GSM, CDMA, and WCDMA.
- **Plug and Play**: The system detects devices automatically, installs required drivers automatically and streamlines user engagement.
- **Simple Architecture:** LTE is more cost-effective for carriers because of its simpler architecture, which reduces Operational Costs (OPEX).

4.3.4. Features of 4G LTE

LTE provides consumers with several benefits, such as the following:

- **Streaming video and audio:** Compared to 2G and 3G, LTE offers quicker upload and download rates.
- **Real-time service connectivity**: Voice over LTE allows customers to communicate with others without any jitter or lag.

- **LTE-Advanced offers even higher rates**: LTE-Advanced offers two to three times quicker upload and download speeds compared to conventional LTE.
- **Carrier aggregation**: Every LTE Advanced device is compatible with regular LTE. By adding bandwidth of up to 100 MHz across five component carriers (bands) with 20 MHz bandwidth each, this LTE-Advanced feature increased network capacity. LTE-A devices enhance signal, speed, and reliability by combining frequencies from several component carriers.

Reduced versions of public LTE networks are known as private LTE networks. They are intended to offer private cellular service over a business's campus, distribution centre, stadiums, airports, and other venues. Unlicensed or shared spectrum is used by private networks to provide coverage for mobile phones and other devices. This covers the unlicensed, worldwide 5 GHz band as well as the 3.5 GHz band, known as the Citizens Broadband Radio Service (CBRS) shared band in the United States. An organization needs an LTE microcell or small cell, core network servers, and compliant devices with a SIM card to set up private LTE services. LTE spectrum bands that can be utilized for private services are supported by a large number of major mobile manufacturers. With the usage of voice over LTE (VoLTE) technology, customers can make phone calls as data packets over the LTE network as opposed to regular voice calls. It may share packets over a network of multiple phone calls and is referred to as packet voice. VoLTE can accommodate a large number of callers and reassign capacity as required. Additional VoLTE features include bandwidth optimization and the ability for the user to determine whether the phone they want to call is available or busy. Before the introduction of LTE, there was no international standard for wireless broadband. GSM had been popular in Asia and Europe before LTE, but CDMA had been adopted by the main mobile operators in other nations, such as the United States and Canada. LTE was designed to bring together a dispersed market and provide network operators with a more effective network.

4.3.5. Major Milestones in LTE's Development

Year	Development
2004	NTT DoCoMo, a Japanese mobile phone operator, proposed making LTE the next international standard for wireless broadband, and work on the LTE standard started.
2006	During a live demonstration, Nokia Networks simultaneously downloaded HD video and uploaded a game via LTE.
2007	Ericsson, a Swedish telecommunications company, demonstrated LTE with a bit rate of 144 Mbps.
2008	Ericsson demonstrated the first LTE end-to-end phone call, and LTE was finalized.
2009	Telia Sonera, a Swedish mobile network operator, made LTE available in Oslo and Stockholm.
2011	LTE-Advanced was finalized in 3GPP Release 10.
2016	3GPP engineers began developing the 5G standard that will eventually succeed LTE.
2017	The first NSA 5G specification was released, becoming widely available in 2018-2019.
2021	5G specification work is ongoing

UNIT SUMMARY

In this module we addressed important concepts to address challenging issues of limited spectrum efficiency and multipath propagations. SS modulation, spectrum is spreaded by modulating the message signal with other sequences for commercial mobile application and thus bringing concept that was originally devised for military communications to now commercial applications We discussed requirements, features and advantages of SS technique. We discussed two primary methods for spreading spectrum namely DSSS and FHSS with necessary details.

As one of the successful multiple access implementations and for increasing wireless network capacity and providing a good alternative to existing time and frequency-based multiple access techniques we discussed CDMA techniques. Types of CDMA channels, its capacity, its call processing etc. and comparison with GSM. Subsequently, we discussed the concept of orthogonality and introduced OFDM *as* multi carrier systems. It also covers *block diagrams of OFDM, its synchronization and efficiency.*

At last, this module discusses the evolution of LTE systems including its working, benefits, its features and major milestones achieved in its journey till date.

KNOW MORE

In today's era, existence of every individual without the mobile or internet services is hard to believe. One can consider different fields starting from health services to fun or entertainment we find penetration of mobile services. Various sectors like banks, hospitals, government or corporate office depend upon mobile applications to their work render their services to citizens. Requirements of better services demand more speed and better customer applications that engage researchers and service providers in upgradation of mobile related services.

Super speeds like 1 Gbps is already being achieved and next on card is 10 Gbps or more. Recent upgradations in 5G technology has promised amplifications of the bandwidth by

1000 times than its predecessor. Admission of 5G network has potential to convert the world into a Wi-Fi zone. Upcoming mobile advancements will also push the usage towards cloud based technologies as new architectures will require reusability and scalability of nodes in the service. In turn, this will enable people to access the network irrespective of places making it ubiquitous. Individuals will be able to access services with same speed on the roads and inside residences. The battery life of mobile device is another area in which tremendous changes are expected. Modern days burning problem is drainage of battery power with increased speed and more consumption of mobile devices is also being addressed. Next generation technology will aim to save the devices power or on the go fast wireless charging in addition to providing high-speed mobile connectivity.

Exercises

Multiple Choice Questions

1. What is the main method by which different users can be distinguished in CDMA?

   a) Timing Allotments
   b) Bands of Frequency
   c) Distinct Codes
   d) Various Schemes of Modulation

2. Which of the following best describes CDMA's main benefit over other multiple access methods?

   a) Increased bandwidth needs
   b) Increased resilience to noise and interference
   c) Easier execution
   d) A set quantity of users

3. Why is the signal being spread across a larger bandwidth than what is necessary in CDMA?

   a) To strengthen the security of signals

b) To cut down on bandwidth usage
c) To make the transmitter design simpler
d) In order to minimize latency

4. What is meant by OFDM?

a) Orthogonal Frequency Division Multiplexing
b) Modulation of the Orthogonal Frequency Domain
c) Multiplexing via Optical Frequency Division
d) Modulation of Data with Orthogonal Frequency

5. In a CDMA system, processing gain (W/R) is 1000, BER is given as 10, what is the maximum number of users a given CDMA system can support?

a) 20 b) 30 c) 1000 d) 100

Answers to Multiple Choice Questions

1. c) 2.b) 3.a) 4.a) 5. d)

Short and Long Answer Type Questions

Short Questions

1.Briefly explain the Spread spectrum. Mention different types of spread spectrum technique.

2. Enlist the main benefits of CDMA.

3. Explain in brief conceptualization of multiple access in CDMA.

4. In CDMA, enlist the properties that must be satisfied by any sequence to be considered as PN sequence.

5. Critically evaluate "CDMA is called interference limited system".

6. Compare Walsh codes vs. PN codes in CDMA.

7. Explain near-far problem and its mitigation in the CDMA system.

8. Explain basic concepts of orthogonality and hence OFDM.

9. Differentiate Single carrier Vs. Multicarrier transmission systems.

10. Explain the working of LTE.
11. Discuss the various features of LTE.

Long Questions

1. Explain direct sequence spread spectrum (DSSS) with necessary diagrams. How is it different from frequency hopping spread spectrum (FHSS)?
2. How can direct sequence spread spectrum (DSSS) strengthen a communication system's resilience and security?
3. Derive the capacity equation of CDMA and compare it with GSM.
4. Enlist types of codes used in CDMA signal generation. Explain any one of them with necessary details.
5. Describe the benefits of frequency hopping spread spectrum (FHSS), including its ability to withstand interference and jamming.

PRACTICAL

1. Design PN sequence for given length and verify its properties.
2. Simulate LTE (Long Term Evolution) downlink PDSCH (Physical DL Shared Channel) with Transmit Diversity using Simulink block-set.

Dynamic QR code for Further Reading

REFERENCES AND SUGGESTED READINGS

1. Theodore Rappaport - Wireless Communications, Principles and Practice-ISBN 0130422320 (Edition) PHI)
2. NPTEL course: https://archive.nptel.ac.in/courses/117/102/117102062/

CO AND PO ATTAINMENT TABLE

Course outcomes (COs) for this course can be mapped with the programme outcomes (POs) after the completion of the course and a correlation can be made for the attainment of POs to analyze the gap. After proper analysis of the gap in the attainment of POs necessary measures can be taken to overcome the gaps.

Table for CO and PO attainment

Course Outcomes	**Expected Mapping with Programme Outcomes** *(1- Weak Correlation; 2- Medium correlation; 3- Strong Correlation)*						
	PO-1	**PO-2**	**PO-3**	**PO-4**	**PO-5**	**PO-6**	**PO-7**
CO-1							
CO-2							
CO-3							
CO-4							
CO-5							

The data filled in the above table can be used for gap analysis.

INDEX

2.5G 17
2G .... xii, xvii, 3, 8, 11, 17, 18, 20, 39, 42, 43, 163
4G .... xii, xvii, xviii, 3, 6, 18, 19, 20, 40, 42, 115, 160, 161, 163
5G .... vi, xii, xvii, 1, 3, 6, 11, 18, 19, 20, 39, 40, 42, 43, 160, 161, 165, 166
AMPS .... xii, 16
Balance property 146
Base Station 3, 36, 99
BPSK 70, 142
capacity .... vi, 1, 22, 25, 37, 39, 41, 110, 112, 114, 115, 119, 126, 129, 150, 160, 164
CCH .... xiv, 3, 10, 24, 34, 43
CDMA standards 138
CDMA, capacity 138
Channel Assignment .... xii, xviii, 25, 39, 44
Co-channel interference .... xviii, 28, 36
Code Division Multiple Access .... xii, 14
Coherence Bandwidth .... xviii, 99, 134
Coherence Time .... xviii, 99, 134
Control Channel 3
Cordless Systems .... xviii, 7
Coverage Area .... xviii, 77
CW 49
Delay Spread .... xviii, 79, 99, 102
diffractionxiv, 54, 55, 62, 66, 67, 68, 69, 71, 83, 86, 91, 93
Diversity .... xv, xvii, xviii, 97, 98, 99, 103, 104, 107, 113, 117, 121, 126, 134, 169
DSSS .... xii, xv, xviii, 138, 140, 141, 142, 143, 144, 152, 166, 169
Equal-Gain Combining .... xii, xviii, 109
Equalization .... xviii, 83, 99, 102, 103
Equivalent Noise Temperature .... xviii, 58
Evolution .... xii, xviii, 1, 2, 11, 16, 18, 39, 43, 160, 169
Fading Margin .... xviii, 86
FDMA .... xii, xiv, 12, 13, 15, 17, 42, 161
FHSS .... xii, xv, xviii, 138, 140, 141, 144, 166, 169
Forward Channel 3
Free Space .... xviii, 49, 50, 70, 91, 92
Free-space path loss 45
Frequency diversity 105, 133, 134
Frequency Division Duplex 4, 162
Frequency Division Duplexing 6
Frequency reuse .... xviii, 22, 28
Full Duplex Systems 4
Fundamentals .... xiv, xviii, 1
GSM .... vii, xii, xviii, 10, 17, 18, 20, 53, 65, 68, 103, 137, 150, 160, 163, 164, 166, 169
Half Duplex Systems 4
handoff .... vi, 1, 26, 27, 40, 41, 72
Handoff .... xii, xiv, xviii, 4, 8, 26, 27, 39, 42, 44
Hata Model .... xviii, 81
HF .... xii, 48, 87
History .... xviii, 1, 2
HPBW .... xii, 48
Indoor propagation .... xviii, 82, 83
Interference .... xii, xviii, 23, 28, 34, 35, 39, 43, 44, 55, 62, 99, 103, 127, 155
large cell 21
Large-Scale .... xviii, 73
Line of Sight .... xii, 47
Link budget 45
Log Normal .... xviii, 74, 76, 91, 92
Log-Distance .... xviii, 74, 91, 93
Longley Rice .... xviii, 79
LTE .... vii, ix, xii, xiii, xv, xviii, 6, 19, 20, 137, 138, 160, 161, 162, 163, 164, 165, 166, 169
Macroscopic diversity 106

Maximal-Ratio Combining........xii, xviii, 108
Maximum length (ML) sequences.. xviii, 147
MIMO....................................vii, xii, xv, xvii, xviii, 19, 20, 97, 98, 115, 116, 117, 119, 120, 121, 122, 123, 124, 126, 127, 128, 129, 130, 131, 132, 134, 135, 161, 162
MIN 10
Mobile Station 4, 99
Mobile Switching Center..................... 4, 40
Multi Path ....xv, xviii, 86
Multi-Carrier................................. xviii, 154
multi-hop 19
Multiple Access Schemes................ xviii, 11
Near-Far Problem .... xviii, 152
Noise figure.. 45
Nyquist... 14
OFDM................vii, ix, xii, xv, xvi, xviii, 19, 20, 130, 137, 138, 154, 156, 157, 158, 159, 161, 166, 168
Okumura .... xviii, 80, 81, 93
Orthogonality................................ xviii, 154
Outage Capacity............................ xviii, 129
Outdoor Propagation....................... xviii, 77
Paging Systems 6
PDC ....xii, 17
Power Control.................xvi, xviii, 152, 153
Propagation models................45, 71, 94, 95
Pseudo noise .... xviii, 146
PSTN. 8
QPSK 70
Quality of Service.......................xii, 12, 162
Radio Systems ....xii, xviii, 1, 5, 18
reflection......................xi, xiv, xviii, 48, 54, 62, 63, 64, 65, 69, 71, 73, 83, 88, 92, 93
Reverse Channel 4
scattering47, 54, 55, 62, 68, 70, 71, 72, 75, 77, 78, 79, 83, 86, 88, 92
sectorization.................................... xviii, 36
Selection Combiningvi, xii, xviii, 97, 107, 132
Shadowing .....xviii, 45, 73, 74, 86, 90, 91, 93
SIM 10, 21, 164
Simplex Systems 4
small cells 21
Small-Scale ....xviii, 72
Smart Antenna ....xvii, xviii, 119, 122
SNR..............................xii, 84, 94, 108, 109, 110, 111, 112, 113, 114, 115, 119, 133, 139, 152, 153, 155, 158
Space diversity..........................xviii, 97, 104, 106, 116, 127, 132
spectrum.............vii, xviii, 12, 21, 22, 23, 25, 27, 33, 41, 49, 56, 57, 59, 79, 80, 105, 106, 115, 116, 121, 122, 130, 132, 137, 138, 139, 140, 142, 143, 144, 154, 155, 156, 159, 162, 163, 164, 166, 168, 169
splitting ....xiv, xviii, 36, 37, 38, 40, 43, 44
Spread Spectrum..............vii, xii, xiii, xv, xviii, 137, 139, 140, 141, 142, 143, 144
Square-Law Combining........... xii, xviii, 110
Statistical Models............................xviii, 88
Street Microcell.....................................xv, 78
Subscriber 4, 10
Switching Type 20
System Capacity...................xviii, 28, 39, 44
Thermal Noise.................................xviii, 55
Time diversity 105, 133, 134
Time Division Duplex..................xiii, 4, 162
Time Division Multiple Access ....xiii, 13, 14
Time-division duplexing 6
Time-invariant.................................xiii, 114
Time-Varying...................................xiii, 114
Traffic Channel 4
Transceiver... 4
UBW ....xiii, 49
VHF 48
Walsh code.....................................xviii, 145
WCDMA 20, 138, 163
Wi-Fi 6, 93, 130, 131, 167
WiMAX ....xiii, 19, 20, 130, 131
wireless communication.......................... 1